Lecture Notes in Mathematics

History of Mathematics Subseries

Volume 2262

Series Editor

Patrick Popescu-Pampu, CNRS, UMR 8524 - Laboratoire Paul Painlevé, Université de Lille, Lille, France

More information about this subseries at http://www.springer.com/series/8909

Robert Penner

Topology and K-Theory

Lectures by Daniel Quillen

With Contribution by Mikhail Kapranov, Kavli Institute
for the Physics and Mathematics of the Universe,
Kashiwa, Chiba, Japan

 Springer

Robert Penner
Institut des Hautes Études Scientifiques
Bures-sur-Yvette, France

ISSN 0075-8434   ISSN 1617-9692 (electronic)
Lecture Notes in Mathematics
ISSN 2193-1771   ISSN 2625-7157 (electronic)
History of Mathematics Subseries
ISBN 978-3-030-43995-8   ISBN 978-3-030-43996-5 (eBook)
https://doi.org/10.1007/978-3-030-43996-5

Mathematics Subject Classification (2010): 19-01, 55-01

This Springer imprint is published by the registered company Springer Nature Switzerland AG
The registered company address is: Gewerbestrasse 11, 6330 Cham, Switzerland

Preface

These are notes from a graduate student course on algebraic topology and K-theory given by Daniel Quillen at the Massachusetts Institute of Technology during 1979–1980. He had just received the Fields Medal for his work on these topics, among others. As a second-semester graduate student myself, it seemed an opportunity to see what all this meant, what a Fields medalist looked and sounded like, what was this exciting new mathematics. Among the gaggle of graduate students, there was also one senior faculty member in attendance, namely Giancarlo Rota.

Dan Quillen was funny and playful with a confident humility from the start. There were points during lectures where he might get stuck and just abandon a proof midstream with a casual *never mind*, and there was always much joking and laughter, enormous energy. A particularly funny moment about which I have periodically chuckled over the years occurred when he was asked some question or other, thought deeply a long moment and answered with a smile that *from a sufficiently enlightened point of view it is obvious*.

Giancarlo was also characteristically funny and playful. He early in the semester asked if we could share class notes, especially when he was absent. At that time in my studies, I was able to scrawl every word uttered in class, and given the attention from Giancarlo, I was driven to revise and legibly copy my notes like never before or since. Furiously taking notes in Sweden a few years later, Peter Jones pulled postdoc me aside and explained that *gentlemen don't take notes*, and I gave up the practice altogether then and there except for an occasional reference or formula.

Giancarlo and I would discuss the lectures which sometimes informed my revisions, and we quickly became friends despite our obviously different circumstances. This was only further cemented a bit less than a decade later when he was a regular Visiting Scholar at the University of Southern California, where I was an assistant professor. We used to joke that we had gone to graduate school together because of the course memorialized here. There were furthermore several occasions from days when I was apparently absent comprising Giancarlo's own notes.

The handwritten notes, mine, ours, and Giancarlo's, are actually in complete sentences with acceptable grammar, and they surfaced recently from a drawer when I moved from my home of thirty years. These are not meant to be polished lecture notes, rather, I have tried to present things as did Quillen, reflected in the hand-written notes, resisting any temptation to change or add notation, details, or elaborations. Indeed, I have been faithful to Quillen's own exposition, even respecting the *board-like presentation* of formulae, diagrams, and proofs, omitting numbering theorems in favor of names and so on. This is meant to be Quillen on Quillen as it happened forty years ago, an informal text for a second-semester graduate student. The intellectual pace of the lectures, namely fast and lively, is Quillen himself, and part of the point here is to capture some of this intimacy. I remember especially the last lecture with its abrupt end, a kind of charmingly embarrassed sayonara, having shared so much of himself during the course.

Despite the avowed goal to present these lectures in their native form, a few insignificant and obvious errors have been corrected. Quillen's own writings are sheer perfection, and the same standard cannot be applied here in the informal context of lecture notes where much ground is covered quickly. The reader is warned, therefore, that there may be small inconsistencies remaining though I and we have done our best in this regard and hope that the overall flow and temperament of these notes might compensate whatever errors may remain.

To be sure, much has happened since then from this categorical perspective started by Grothendieck, and I am grateful to my friend Misha Kapranov for contributing an Afterword to this volume in order to make it more useful to current students and also for refining and correcting my own notes. It is likewise a pleasure to thank Cécile Gourgues for her superb transcription from the handwritten notes to beautiful LaTeX and the Institut des Hautes Études Scientifiques for supporting this project. Thanks also to Geoff Taylor and Tony Philp for their hospitality in Fiji, where this manuscript was ultimately completed. Let me finally dedicate this little volume to the memories of my friend Gianco and my love Lexy.

Savusavu, Fiji   Robert Penner
November 2018

Contents

1   Group Extensions and Cohomology   1

2   Categories and Their Nerves.............................   5

3   Simplicial Objects   9

4   Normalization and Conical Contractibility   15

5   Effaceable δ-Functors   21

6   (Co)homology of Cyclic Groups   25

7   An Application to the Schur–Zassenhaus Theorem.............   33

8   The Yoneda Lemma   39

9   Kan Formulae...   45

10 Abelian and Additive Categories...........................   51

11 Diagram Chasing in Abelian Categories   57

12 Fibered and Cofibered Categories..........................   61

13 Examples of Fibered Categories   67

14 Projective Resolutions   71

15 Analogues of Homotopy Liftings...........................   75

16 The Mapping Cylinder and Mapping Cone   83

17 Derived Categories   89

18 The First Homotopy Property.............................   95

19 Group Completions and Grothendieck Groups...............   101

20 Devissage and Resolution Theorems   105

21   Exact Sequences of Homotopy Classes 109

22   Spectral Sequences 113

23   Spectral Sequences Continued 119

24   Hyper-Homology Spectral Sequences 125

25   Generalized Kan Formulae 131

26   The Hochschild–Serre Spectral Sequence 137

27   Resolution for Exact Categories 143

28   $K_0 A \cong K_0 A[T]$ 149

29   Classifying Spaces 155

30   Higher K-Groups 159

31   The Category $Q\mathcal{M}$ 165

32   Homotopy Equivalence 169

33   A Filtration of $Q(\mathcal{P}_A)$ 175

34   Bi-simplicial Sets and Dold–Thom 179

35   Homology of $Q(\mathcal{P}_A)$ and the Tits Complex ... 185

36   Long Exact Sequences of K-Groups 189

37   Localization 195

38   The Plus Construction, K_1 and K_2 201

Afterword by Mikhail Karpranov 207

References 209

Index 211

Chapter 1
Group Extensions and Cohomology

Consider an *extension* E *of* the group G *by* the group N

$$* \hookrightarrow N \xrightarrow{i} E \xrightarrow{p} G \longrightarrow *,$$

so $i : N \hookrightarrow E$ is an injection, $p : E \twoheadrightarrow G$ a surjection and the kernel $\mathrm{Ker}\, p$ of p equals the image $\mathrm{Im}\, i$ of i, i.e., a short exact sequence. The obvious notion of iso-morphisms of extensions is given by a commutative diagram

and we let $\mathcal{E}(G, N)$ denote the collection of isomorphism classes.

If $u : G' \to G$ is a homomorphism, then there is a pull back

$$u^* : \mathcal{E}(G, N) \to \mathcal{E}(G', N)$$

defined by the diagram

where $E \times_G G' = \{(a, b) : a \in E, b \in G' \text{ and } pa = ub\}$ and π_1 is induced by pro-jection onto the first factor.

© Springer Nature Switzerland AG 2020

R. Penner, *Topology and K-Theory*, Lecture Notes in Mathematics 2262,

https://doi.org/10.1007/978-3-030-43996-5_1

The pull back of an epimorphism is again an epimorphism, the pull back preserves fibers, i.e., cosets of N, and we can check directly that the induced map $N \to N$ in the diagram is the identity map.

Given an extension $* \longrightarrow N \xrightarrow{i} E \xrightarrow{p} G \longrightarrow *$ and given $g \in G$, choose some $h \in E$ so that $p(h) = g$. Then h acts on E by conjugation and in particular on N since $N \lhd E$. Thus we have

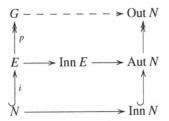

where Aut denotes the automorphism group, Inn $\lhd$ Aut the inner automorphism group, and the quotient Out $=$ Aut/Inn is the outer automorphism group. In particular for N abelian, we have Aut $N =$ Out N and get a homomorphism

$$G \longrightarrow \text{Aut } N$$
$$g \longmapsto (n \mapsto i^{-1}(h\, i(n)\, h^{-1}))$$

where $h \in E$ projects to $g = p(h)$.

Thus for N abelian **as we assume from now on,** N is a G-module and

$$\mathcal{E}(G, N) = \coprod_{G \xrightarrow{\theta} \text{Aut}N} \mathcal{E}_\theta(G, N),$$

where $\mathcal{E}_\theta(G, N)$ denotes the extensions compatible with the homomorphism $\theta : G \to$ Aut N. $\mathcal{E}_\theta(G, N)$ is covariant in the G-module N for G-equivariant maps in the sense that given two G-actions θ on N and θ' on N', a G-equivariant $u : N \to N'$ induces $\mathcal{E}_\theta(G, N) \to \mathcal{E}_{\theta'}(g, N')$, i.e.,

$$
\begin{array}{ccccccccc}
* & \longrightarrow & N & \longrightarrow & E & \longrightarrow & G & \longrightarrow & * \\
& & \downarrow{\scriptstyle u} & & \downarrow & & \downarrow & & \\
* & \longrightarrow & N' & \longrightarrow & E' & \longrightarrow & G & \longrightarrow & *
\end{array}
$$

where $E' = N' \times^N E = N' \times E$ modulo the action of N, i.e., $(n'n, e) \sim (n', ne)$. Carry on to define multiplication in E' by

$$(c\ell(n'_1, e_1)) \cdot (c\ell(n'_2, e_2)) \stackrel{d}{=} (c\ell(n'_1 \cdot p(e_1)\, n'_2, e_1\, e_2)),$$

where $c\ell$ means equivalence class, the motivation being

$$n_1' \, e_1 \, n_2' \, e_2 = n_1' \, e_1 \, n_2' \, e_1^{-1} \, e_1 \, e_2,$$

and we check that this is well-defined.

Description by Means of Cocycles

Given the extension $* \longrightarrow N \xrightarrow{i} E \xrightarrow{p} G \longrightarrow *$, choose a section $s : G \to E$ of p, that is, a map of sets so that $ps = \mathrm{id}_G$. Then every element of E can be uniquely written as

$$i(u)\, s(g), \quad \text{for } n \in N \text{ and } g \in G.$$

The multiplication is

$$[i(n_1)\, s(g_1)]\, [i(n_2)\, s(g_2)] \overset{d}{=} i(n_1 + g_1\, n_2)\, s(g_1)\, s(g_2).$$

The notation is that N is additive, E and G multiplicative, and we denote the action of g on N multiplicatively, so the product can be written

$$= i(n_1 + g_1 n_2)\, i(f(g_1, g_2))\, s(g_1\, g_2)$$
$$= i(n_1 + g_1\, n_2 + f(g_1, g_2))\, s(g_1\, g_2),$$

where $f : G \times G \to N$ depending on the choice of s.

Conclusion the group operation on E is determined by f where

$$s(g_1)\, s(g_2) = i(f(g_1, g_2))\, s(g_1\, g_2).$$

Exercise 1.1 Associativity in E implies

$(*_1)$ $\qquad g_1\, f(g_2, g_3) - f(g_1\, g_2, g_3) + f(g_1, g_2\, g_3) - f(g_1, g_2) = 0.$

Hint: $(s(g_1)\, s(g_2))\, s(g_3)$ gives the second and fourth terms, and $s(g_1)\, (s(g_2)\, s(g_3))$ gives the first and third.

Such an $f : G \times G \to N$ is called a *2-cocycle of G with values in the G-module* N.

Exercise 1.2 Suppose $\tilde{s} : G \to E$ is another section of p. Then we get $h : G \to N$ by $\tilde{s}(g) = i(h(g))\, s(g)$. Show

$(*_2)$ $\qquad \tilde{f}(g_1, g_2) - f(g_1, g_2) = g_1\, h(g_2) - h(g_1\, g_2) + h(g_1).$

Hint: $\tilde{s}(g_1)\, \tilde{s}(g_2) = i\, h(g_1)\, s(g_2)\, i\, h(g_2)\, s(g_2).$

The right-hand side of ($*_2$) is called the 1-*coboundary of h.*

Conclusion Can attach to $* \to N \to E \to G \to *$, for N a G-module, a well-defined element of

$$H^2(G, N) \stackrel{d}{=} \text{(group of 2-cocycles with values in } N)/\text{(group of 1-coboundaries)}.$$

An element of the denominator is a 2-cocycle of the form

$$(g_1, g_2) \mapsto g_1 h(g_2) - h(g_1 g_2) + h(g_1).$$

Theorem *In the above way, the isomorphism classes of extensions of G by the G-module N are in $1 - 1$ correspondence with elements of $H^2(G, N)$.*

Proof A messy but routine exercise.   □

General Formulae for the Cochain Complex of G with Values in the G-Module N

Consider the cochain complex

$$\cdots \longrightarrow 0 \longrightarrow C^0(G, N) \stackrel{\delta}{\to} C^1(G, N) \stackrel{\delta}{\to} C^2(G, N) \stackrel{\delta}{\to} \cdots,$$

with $C^q(G, N) = \text{Maps}(G^q, N)$, where G^q is the q-fold Cartesian product and $\text{Maps}(X, Y)$ denotes all set maps from X to Y with $C^0(G, N) = N$, and for $f \in C^q$

$$(\delta f)(g_1, \ldots, g_{q+1}) \stackrel{d}{=} g_1 f(g_2, \ldots, g_{q+1}) - f(g_1 g_2, g_3, \ldots, g_{q+1})$$
$$+ f(g_1, g_2 g_3, g_4, \ldots, g_{q+1}) - \cdots \pm f(g_1, \ldots, g_q).$$

A routine but messy exercise which we shall explicate later shows that $\delta^2 = 0$, and we define

$$H^q(G, N) = \text{(Kernel of } \delta \text{ on } C^q)/\text{(Image of } C^{q-1} \text{ under } \delta).$$

To compute H^0, if $n \in C^0(G, N) \equiv N$, then $(\delta^0 n)(g) = gn - n$, and so

$$H^0(G, N) = \{n : gn = n \text{ for all } g \in G\}$$
$$= \text{subgroup of elements of } N \text{ fixed by } G$$
$$= N^G.$$

And by the Theorem, $H^2(G, N)$ is the collection of isomorphism classes of extensions of G by N.

Chapter 2
Categories and Their Nerves

Let us next compute $H^1(G, N)$ for N a G-module.
$h \in Z^1(G, N) =$ cycles in $C^1 = \mathrm{Ker}(\delta : C^1 \to C^2)$ means

$$h(g_1 \, g_2) = g_1 \, h(g_2) + h(g_1) \,,$$

and h is called either a *derivation* or a *crossed homomorphism* in this case. They can be interpreted as follows:
    Let θ be an automorphism of an extension E of G by N, so that

$$
\begin{array}{ccccccccc}
* & \longrightarrow & N & \xrightarrow{\ i\ } & E & \xrightarrow{\ p\ } & G & \longrightarrow & * \\
 & & \big\| & & \big\downarrow{\scriptstyle \theta} & & \big\| & & \\
* & \longrightarrow & N & \xrightarrow{\ i\ } & E & \xrightarrow{\ p\ } & G & \longrightarrow & *
\end{array}
$$

Consider $E \to E$ where $e \mapsto \theta(e)e^{-1}$ so $p(\theta(e)e^{-1}) = 1$ while

$$
\begin{aligned}
\theta(e \cdot i(n))(e \cdot i(n))^{-1} &= \theta(e)\theta(i(n)) \, i(n)^{-1} \, e^{-1} \\
&= \theta(e) \, e^{-1}.
\end{aligned}
$$

So this map is constant on cosets of N, and there is some $h : G \to N$ so that

$$\theta(e) \, e^{-1} = i(h(pe)) \,.$$

We claim that this h is a derivation.
    To see this, take $g_1, g_2 \in G$, lift to $e_1, e_2 \in E$ and compare the two sides of

$$
\begin{aligned}
\theta(e_1 \, e_2) &= \theta(e_1) \, \theta(e_2) \\
&= i \, h(g_1) \, e_1 \, i \, h(g_2) \, e_2 \,.
\end{aligned}
$$

The left-hand side is $i(h(g_1 \, g_2)) \, e_1 \, e_2$, while the right-hand side is

© Springer Nature Switzerland AG 2020
R. Penner, *Topology and K-Theory*, Lecture Notes in Mathematics 2262,
https://doi.org/10.1007/978-3-030-43996-5_2

$i\,h(g_1)\,i(g_1\,h(g_2))\,e_1\,e_2 = i(h(g_1) + g_1\,h(g_2))\,e_1\,e_2$, as desired.

Conversely, defining $\theta(e) = i\,h(pe)\,e$ gives an automorphism.

Thus $Z^1(G, N) =$ group of automorphisms of **any** extension of G by N.

Exercise $B^1(G, N) =$ coboundaries in $C^1 = \operatorname{Im}\{\delta : C^0 \to C^1\}$ is the subgroup consisting of all inner automorphisms by elements of N.

Note that for G acting trivially on N, we have $\delta_0 \equiv 0$, and a derivation is just a homomorphism, so $H^0(G, N) = N$, $H^1(G, N) =$ group homomorphisms from G to N and $H^2(G, N) =$ isomorphism classes of central extensions of G by the abelian group N.

Categories

A *category* $\mathcal{C}$ consists of a class $\operatorname{Ob}\mathcal{C}$ of *objects*, and for $X, Y \in \operatorname{Ob}\mathcal{C}$ we are given a **set** $\operatorname{Hom}_{\mathcal{C}}(X, Y)$ of *maps* or *morphisms*, and for $X, Y, Z \in \operatorname{Ob}\mathcal{C}$ a *composition map*

$$\operatorname{Hom}_{\mathcal{C}}(X, Y) \times \operatorname{Hom}_{\mathcal{C}}(Y, Z) \longrightarrow \operatorname{Hom}_{\mathcal{C}}(X, Z)$$
$$f \times g \longmapsto gf$$

If $f \in \operatorname{Hom}_{\mathcal{C}}(X, Y)$, then we sometimes write simply $f : X \to Y$ and may call f an *arrow* in $\mathcal{C}$.

Axioms (1) associativity of composition,
(2) existence of identity maps,
(3) the sets $\operatorname{Hom}_{\mathcal{C}}(X, Y)$ are disjoint as X, Y varies in $\operatorname{Ob}\mathcal{C}$.

A *small category* is a category $\mathcal{C}$ so that $\operatorname{Ob}\mathcal{C}$ is a set.

Example 2.1 Given a partially ordered set $(I, \leq)$, called a *poset* for short, we get a small category $\tilde{I}$ so that $\operatorname{Ob}\tilde{I} = I$ and

$$\operatorname{Hom}_{\tilde{I}}(X, Y) = \begin{cases} \{(x, y)\} \text{ (a 1-element set)}, & \text{if } x \leq y, \\ \varnothing, & \text{if } (x \leq y) \text{ does not hold.} \end{cases}$$

Example 2.2 For a monoid M, so the operation is associative and there are identity maps, we get a category $\tilde{M}$ with

$$\operatorname{Ob}\tilde{M} = \{*\}, \qquad \operatorname{Hom}_{\tilde{M}}(*, *) = M,$$

where composition is multiplication in M.

In a category $\mathcal{C}$, $f : X \to Y$ is an *isomorphism* if there is

$$g : Y \to X \text{ so that } gf = \operatorname{id}_X \text{ and } fg = \operatorname{id}_Y.$$

A *groupoid* is a category in which every morphism is an isomorphism. Note that $\tilde{I}$ is groupoid if and only if all elements are incomparable. Also $m \in \tilde{M}$ is an isomorphism if m^{-1} exists, and $\tilde{M}$ is a groupoid if and only if M is a group.

Let C be a small category. Let $N_0(C)$ be the set of objects and

$$N_1(C) = \bigcup_{X,Y \in \mathrm{Ob}\,C} \mathrm{Hom}_C(X, Y).$$

We have maps

$$N_1(C) \underset{\text{target}}{\overset{\text{source}}{\rightrightarrows}} N_0(C).$$

identity map

Let now

$$N_q(C) = \text{diagrams in } C \text{ of composable arrows of length } q$$

and call an element of $N_q(C)$ a *q-simplex*. We have maps

$$N_q(C) \xrightarrow{\ d_j,\ j=0,\ldots,q\ } N_{q-1}(C)$$

where

$$d_j\left(X_0 \leftarrow \cdots \xleftarrow{f_j} X_j \xleftarrow{f_{j+1}} X_{j+1} \cdots \leftarrow X_q\right)$$
$$= \left(X_0 \leftarrow \cdots \xleftarrow{f_{j-1}} X_{j-1} \xleftarrow{f_j f_{j+1}} X_{j+1} \leftarrow \cdots X_q\right).$$

We also have maps

$$N_{q-1}(C) \xrightarrow{\ s_j,\ j=0,\ldots,q-1\ } N_q(C)$$

where

$$s_j(X_0 \leftarrow \cdots \leftarrow X_j \cdots \leftarrow X_{q-1}) = \left(X_0 \leftarrow \cdots \leftarrow X_j \xleftarrow{\text{id}} X_j \leftarrow \cdots X_q\right).$$

This so-called simplicial set is the *nerve of the category* C.

A functor $F : C_1 \to C_2$ induces a map

$$N(C_1) \longrightarrow N(C_2).$$

Consider the following posets.

$[q]^{\mathrm{op}} = \{0, 1, \ldots, q\}$ with the **wrong** order. A functor $\widetilde{[q]^{\mathrm{op}}} \to C$ is the same thing as a q-simplex in C, where $\widetilde{[q]^{\mathrm{op}}}$ is the category of Example 1 derived from the poset $\widetilde{[q]^{\mathrm{op}}}$.

An order preserving map $[p]^{op} \xrightarrow{\theta} [q]^{op}$ will induce a map

$$N_q(\mathcal{C}) \xrightarrow{\theta^*} N_p(\mathcal{C}),$$

and any θ^* has a canonical presentation

$$s_{i_1} \ldots s_{i_p} d_{j_1} \ldots d_{j_k}$$

where

$$i_1 > \ldots > i_p \text{ and } j_1 < \ldots < j_k$$

using the commutation relations.

Chapter 3
Simplicial Objects

Let $\triangle$ be the small category with objects $\mathrm{Ob}\ \triangle = \{[p]|p \geq 0\}$, where $[p] = \{0, \ldots, p\}$ with the usual order and $\mathrm{Hom}\ \triangle$ the set of order-preserving maps.

If $\mathcal{C}$ is any category, then a *simplicial object* in $\mathcal{C}$ is a contravariant functor $X : \triangle \to \mathcal{C}$. In particular the nerve of a small category $\mathcal{C}$ is such, and the nerve itself is a functor on $\triangle$.

Morphisms in $\triangle$

There is a nice system of generators as follows

$$faces\ \partial_i :\ [p-1] \to [p],$$
$$\text{the unique order-preserving injective}$$
$$\text{map with } i \text{ omitted from the image,}$$

$$degeneracies\ \sigma_j :\ [p+1] \to [p],$$
$$\text{the unique order-preserving surjective}$$
$$\text{map identifying } j \text{ and } j+1.$$

Now, any injective map $\varepsilon : [p] \to [q]$ has a canonical factorization

$$\varepsilon = \partial_{i_r} \ldots \partial_{i_1},$$

where $i_1 < \ldots < i_r$ are the elements omitted in the image of ε. Likewise any surjective map $\eta : [p] \to [q]$ has a canonical factorization

$$\eta = \sigma_{j_1} \ldots \sigma_{j_s},$$

where $j_1 < \ldots < j_s$ are the elements so that $\eta(j) = \eta(j+1)$.

© Springer Nature Switzerland AG 2020
R. Penner, *Topology and K-Theory*, Lecture Notes in Mathematics 2262,
https://doi.org/10.1007/978-3-030-43996-5_3

Thus any map θ in $\triangle$ factors as

$$\theta = \partial_{i_r} \cdots \partial_{i_1} \sigma_{j_1} \cdots \sigma_{j_s}.$$

Axioms for *simplicial objects* are (essentially) rules for composition:

$$a \leq b \quad \text{implies} \quad \partial_a \partial_b = \partial_{b+1} \partial_a,$$
$$a \leq b \quad \text{implies} \quad \sigma_b \sigma_a = \sigma_a \sigma_{b+1},$$
$$\text{for all } a, b, \quad \sigma_a \partial_b = \begin{cases} \text{id}, & \text{if } a \in \{b, b-1\}, \\ \partial_b \sigma_{a-1}, & \text{if } \quad a > b, \\ \partial_{b-1} \sigma_a, & \text{if } \quad a < b-1. \end{cases}$$

Proposition *A simplicial object* $X : \triangle \to C$ *in a category* C *is a collection*

$$X_p \in \mathrm{Ob}\, C, \quad \text{for } p \geq 0,$$

together with morphisms

$$d_i : X_p \longrightarrow X_{p-1}, \text{ for } i = 0, \ldots, p,$$
$$s_i : X_p \longrightarrow X_{p+1}, \text{ for } i = 0, \ldots, p,$$

so that

$$a \leq b \quad \text{implies} \quad d_b d_a = d_a d_{b+1},$$
$$a \leq b \quad \text{implies} \quad s_a s_b = s_{b+1} s_a,$$
$$\text{for all } a, b, \quad d_b s_a = \begin{cases} \text{id}, & \text{for } b \in \{a, a+1\}, \\ s_{a-1} d_b, & \text{for } a > b, \\ s_a d_{b-1}, & \text{for } a < b-1. \end{cases}$$

Given a morphism $\theta : [p] \longrightarrow [q]$, *there is an induced* $\theta^* : X_p \longleftarrow X_q$, *where* θ^* *is a composition.*

$$\theta^* = \underbrace{s \ldots s}_{\text{indices go down}} \quad \underbrace{d \ldots d}_{\text{indices go up}}.$$

Chains

Let C be a small category and let $F : C \to \mathrm{Ab}$ be a covariant functor to the category Ab of abelian groups. We shall presently define a *simplicial abelian group*. A motivating example is described by the diagram

$$\cdots \quad \bigoplus_{X_0 \leftarrow X_1 \leftarrow X_2} F(X_2) \Longrightarrow \bigoplus_{X_0 \leftarrow X_1} F(X_1) \Longrightarrow \bigoplus_{X_0} F(X_0).$$

Precisely with $\coprod = \oplus$ the direct sum, define

$$C_p(\mathcal{C}, F) \overset{d}{=} \coprod_{\substack{X_0 \leftarrow \dots \leftarrow X_p \\ \in N_p(\mathcal{C})}} F(X_p) \, .$$

Given a monotone $\theta : [p] \to [q]$, define

$$\theta^* : C_q(\mathcal{C}, F) \longrightarrow C_p(\mathcal{C}, F)$$

as follows. Suppose α is in the summand belonging to

$$X_0 \longleftarrow \cdots \longleftarrow X_q$$

and let $X_j \xrightarrow{\ \text{in}_j\ } \coprod_j X_j$ be the canonical map. Then

$$\alpha = \text{in}_{(X_0 \leftarrow \dots \leftarrow X_q)} \xi \, , \quad \text{for} \quad \xi \in F(X_q) \, ,$$

and we have

$$\theta^*(X_0 \longleftarrow \cdots X_{\theta(p)} \cdots \longleftarrow X_q) = X_{\theta(0)} \longleftarrow \cdots \longleftarrow X_{\theta(p)} \, .$$

Then there is a map $X_q \xrightarrow{\ u\ } X_{\theta(p)}$ as part of this simplex, and we define

$$\theta^* \, \text{in}_{(X_0 \leftarrow \dots \leftarrow X_q)} \, \alpha = \text{in}_{\theta^*(X_0 \leftarrow \dots \leftarrow X_q)} F(u)(\alpha) \, .$$

Exercise If F as above is contravariant, then modify the construction to make sense of

$$\cdots \coprod_{X_0 \leftarrow X_1 \leftarrow X_2} F(X_0) \rightrightarrows \coprod_{X_0 \leftarrow X_1} F(X_0) \rightrightarrows \coprod_{X_0} F(X_0) \, .$$

A *co-simplicial object* in $\mathcal{C}$ is a covariant functor $\Delta \to \mathcal{C}$.

Co-chains
Suppose $F : \mathcal{C} \to \text{Ab}$ is covariant. Construct

$$\prod_{X_0} F(X_0) \underset{\partial_1}{\overset{\partial_0}{\rightrightarrows}} \prod_{X_0 \leftarrow X_1} F(X_0) \rightrightarrows \prod_{X_0 \leftarrow X_1 \leftarrow X_2} F(X_0) \cdots$$

as before, namely

$$C^p(\mathcal{C}, F) = \prod_{X_0 \leftarrow \dots \leftarrow X_p \in N_p \mathcal{C}} F(X_0) \, ,$$

and given $\theta : [p] \to [q]$, define $\theta_* : C^p(\mathcal{C}, F) \to C^q(\mathcal{C}, F)$, for $\alpha \in \prod_{N_p \mathcal{C}} F(X_0)$, as follows

$$\mathrm{pr}_{X_0 \leftarrow \cdots \leftarrow X_q} \, \theta_* \, \alpha \longmapsto F(u) \, \mathrm{pr}_{\theta^*(X_0 \leftarrow \cdots \leftarrow X_q)} \, \alpha \, ,$$

where $\mathrm{pr}_{(X_0 \leftarrow \cdots \leftarrow X_q)}$ denotes the canonical map.

Let now

$$\alpha \in \prod_{X_0 \leftarrow \cdots \leftarrow X_p \in N_p \, \mathcal{C}} F(X_0) \in C^p(\mathcal{C}, F)$$

and write $\alpha(X_0 \leftarrow \cdots \leftarrow X_p)$ for $\mathrm{pr}_{X_0 \leftarrow \cdots \leftarrow X_p} \alpha$, so given $\partial_j : [p] \to [p+1]$, we have

$$(\partial_j \alpha)(X_0 \longleftarrow \cdots \longleftarrow X_{p+1}) = \begin{cases} \alpha (X_0 \longleftarrow \cdots \widehat{X}_j \longleftarrow \cdots \longleftarrow X_{p+1}), & \text{for } j \neq 0, \\ F(u) \, \alpha(X_1 \longleftarrow \cdots \longleftarrow X_{p+1}), & \text{for } j = 0. \end{cases}$$

Suppose now given a simplicial abelian group

$$\cdots \ C^3 \Rrightarrow C_2 \Rightarrow C_1 \Rightarrow C_0 \, ,$$

and define

$$d = \sum_{j=0}^{p-1} (-1)^j \, d_j : C_p \longrightarrow C_{p-1} \, .$$

Claim $d^2 = 0$.

Proof For $j \geq i$, we have $d_j \, d_i = d_i \, d_{j+1}$, and

$$\begin{aligned} d^2 &= \sum_{j=0}^{p-1} (-1)^j \, d_j \sum_{k=0}^{p} (-1)^k \, d_k \\ &= \sum_{\substack{0 \leq j \leq p-1 \\ 0 \leq k \leq p}} (-1)^{j+k} \, d_j \, d_k \\ &= \sum_{0 \leq j < k \leq p} (-1)^{j+k} \, d_j \, d_k + \underbrace{\sum_{\substack{0 \leq k \leq j \leq p-1 \\ \quad \overset{\shortparallel}{d_k \, d_{j+1}}}} (-1)^{j+k} \, d_j \, d_k}_{= \sum_{\substack{0 \leq a \leq p-1 \\ a < b \leq p}} (-1)^{a+b-1} \, d_a \, d_b} \\ &= 0 \, . \end{aligned}$$

$\square$

The homology $H_*(C_p(C, F), d)$ of $\mathcal{C}$ with values in F is thus defined, and the cohomology $H_*(C^p(\mathcal{C}, F), \delta)$ of $\mathcal{C}$ with values in F is now defined as above, where

$$\delta = \sum_{i=0}^{p+1}(-1)^i\,\partial_i\,.$$

For $\alpha \in C^p(\mathcal{C}, F)$, we have $\alpha(X_0 \leftarrow \cdots \leftarrow X_p) \in F(X_0)$ and

$$(\delta\,\alpha)\left(X_0 \underset{u}{\leftarrow} \cdots X_j \leftarrow \cdots X_{p+1}\right)$$

$$= u_*\,\alpha(X_1 \leftarrow \cdots \leftarrow X_{p+1}) - \alpha(X_0 \leftarrow \widehat{X_1} \leftarrow \cdots \leftarrow X_{p+1})$$
$$+ \alpha(X_0 \leftarrow X_1 \leftarrow \widehat{X_2} \leftarrow \cdots \leftarrow X_{p+1}) - \cdots.$$

Now, if G is a group and N a G-module, then interpret N as a functor

$$\tilde{N} : \tilde{G} \longrightarrow \text{Ab}$$
$$* \underset{g}{\to} * \longmapsto N \underset{g}{\to} N\,,$$

so $C^p(\tilde{G}, \tilde{N}) = \prod_{N_p(\tilde{G})} N$, and an element of $N_p(G)$ is

$$* \xleftarrow{g_1} * \xleftarrow{g_2} \cdots \xleftarrow{g_p} *\,.$$
$$\ \ 0 \qquad 1 \qquad\qquad p$$

Therefore $C^p(\tilde{G}, \tilde{N}) = \text{Maps}(G^p, N)$, as before and

$$(\delta\,\alpha)(g_1, \ldots, g_p) = g_1\,\alpha(g_2, \cdots, g_p)$$
$$- \alpha(g_1\,g_2, g_3, \ldots, g_p)$$
$$+ \cdots$$

as before and hence $H^p(\tilde{G}, \tilde{N}) = H^p(G, N)$.

Chapter 4
Normalization and Conical Contractibility

Let I be a poset (short for partially ordered set) and take $\widetilde{I}$ as before. $N_p(\widetilde{I}) = \{$chains of sequences$\}$, i.e., $X_0, \ldots, X_p \in I$ with $X_0 \geq \ldots \geq X_p$. When the weak equality is actually equality, then we have a degeneracy, and by collapsing we can get a non-degenerate simplex.

Exercise Suppose that $X = \{X_p\}$, a simplicial set. Show any simplex can be uniquely expressed as $x = \eta^* y$ where η is a surjective monotone map $\eta : [p] \to [q]$ and y is non-degenerate in the sense it is not in the image of any degeneracy.

Recall the definition of $H_*(\widetilde{I}, F)$ for a functor $F : \widetilde{I} \to \mathrm{Ab}$.

Take F to be a constant functor $X \mapsto A$ and $f \mapsto \mathrm{id}_A$ for all morphisms f. Taking sums in $C_*(\widetilde{I}, F)$ only over non-degenerate simplexes of any poset gives rise to a *simplicial complex* in the sense of combinatorial topology.

Vertices are elements of I and non-degenerate simplexes are $X_0 > \ldots > X_p$.

If K is a simplicial complex, then let I be the poset of simplexes in K. Then the simplicial complex associated to I is the barycentric sub-division of K.

Note that the usual complex of chains on the simplicial complex belonging to I is the non-degenerate part of $C_*(\widetilde{I}, A)$.

Reference Dold–Puppe, Ann. Inst. Fourier (1961).

Normalization Theorem Let $\{C_p\}$ be a simplicial abelian **group**. Then

$$C_p = \sum_{j=0}^{p-1} \mathrm{Im}\,\{s_j : C_{p-1} \longrightarrow C_p\}$$

$$\oplus \bigcap_{j=1}^{p} \mathrm{Ker}\,\{d_j : C_p \longrightarrow C_{p-1}\}$$

is a decomposition of the complex $\cdots \to C_{p+1} \xrightarrow{d} C_p \xrightarrow{d} C_{p-1} \to \cdots$. Moreover calling the first summand $C_p^{\deg}$, the *degenerate sub-complex*, $\{C_p^{\deg}\}$ has trivial homology.

© Springer Nature Switzerland AG 2020

R. Penner, *Topology and K-Theory*, Lecture Notes in Mathematics 2262,

https://doi.org/10.1007/978-3-030-43996-5_4

Consequence

$$C_* \longrightarrow C_*/C_*^{\text{deg}}$$

induces a homology isomorphism and in particular

$$
\begin{array}{ccc}
C_p(\widetilde{I}, A) & \longrightarrow & C_p(\widetilde{I}, A)/C_p(\widetilde{I}, A)^{\text{deg}} \\
\| & & \| \\
\coprod_{x_0 \geq \ldots \geq x_p} A & \overset{\text{homology}}{\underset{\text{isomorphism}}{\dashrightarrow}} & \coprod_{x_0 > \ldots > x_p} A
\end{array}
$$

Conclusion For a constant functor A, $H_*(\widetilde{I}, A)$ is the same as the homology of the simplicial complex, and likewise for cohomology.

Example 4.1 $[p]$ as a simplicial complex is $\triangle(p)$ so that

$$H_*([p], A) = \begin{cases} A, & * = 0, \\ 0, & \text{else.} \end{cases}$$

Example 4.2 The poset described by the Hasse diagram

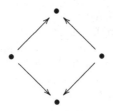

has the "same" simplicial complex so

$$H_* = \begin{cases} A, & * = 0, 1, \\ 0, & \text{else.} \end{cases}$$

Exercise Suppose $\{C_p\}$ is the simplicial abelian group

$$\ldots C_2 \Rrightarrow C_1 \rightrightarrows C_0$$

and suppose there is $s_{-1} : C_p \to C_{p+1}$, for all $p \geq 0$, so that the usual identities hold. Then

$$H_* C = 0 \text{ for } * > 0.$$

For example, given a simplicial abelian group

$$A_3 \Rrightarrow A_2 \Rightarrow A_1 \rightrightarrows A_0 \, ,$$

define

$$C_p = A_{p+1} \text{ with } d_j, s_j \text{ for } C \text{ given respectively by } d_{j+1}, s_{j+1} \text{ for } A \, .$$

Then $\{C_p\}$ is a simplicial abelian group and s_{-1} exists as in the exercise, so then $H_* C = 0$.

Why is $H_1 C = 0$? Well, we have

$$C_2 \xrightarrow{d} C_1 \xrightarrow{d} C_0$$

with

$$d \, s_{-1} = (d_0 - d_1 + d_2) \, s_{-1} = \text{id} - s_{-1} \, d_0 + s_{-1} \, d_1,$$

and

$$s_{-1} \, d = s_{-1}(d_0 - d_1) = s_{-1} \, d_0 - s_{-1} \, d_1,$$

so that

$$d \, s_{-1} + s_{-1} \, d = \text{id} \text{ which implies } H_1 C = 0 \, .$$

Exercise Generalize this. In general, the extra degeneracy map gives a chain homotopy id $\simeq 0$, and C is said to be *conically contractible*.

Application Suppose that $\mathcal{C}$ is a small category and Y a fixed object of $\mathcal{C}$. Take $F(X) = \coprod\limits_{X \leftarrow Y} A$ for A an object in Ab. Then $C_*(\mathcal{C}, F)$ is

$$\coprod\limits_{X_0 \leftarrow X_1 \leftarrow X_2} \coprod\limits_{X_2 \leftarrow Y} A \Rrightarrow \coprod\limits_{X_0 \leftarrow X_1} \coprod\limits_{X_1 \leftarrow Y} A \rightrightarrows \coprod\limits_{X_0 \leftarrow Y} A = C_0(\mathcal{C}, F),$$

so we have

$$\coprod\limits_{X_0 \leftarrow X_1 \leftarrow X_2 \leftarrow Y} A \Rrightarrow \coprod\limits_{X_0 \leftarrow X_1 \leftarrow Y} A \overset{d_0}{\underset{d_1}{\rightrightarrows}} \coprod\limits_{X_0 \leftarrow Y} A \, .$$

We get an extra degeneracy map, where we have s_{p+1} here, so acyclic as above since conically contractible.

Note that the direct sum of this construction over all Y gives the construction of the previous exercise. Also note $C \hookrightarrow C'$ together with $H_* C' = 0$ implies $H_* C = 0$.

Cohomology

Consider $C^*(\mathcal{C}, F)$ with F covariant

$$\prod_{X_0} F(X_0) \underset{\partial_1}{\overset{\partial_0}{\rightrightarrows}} \prod_{X_0 \leftarrow X_1} F(X_0) \rightrightarrows \cdots$$

and take for fixed Y, $F(X) = \prod_{Y \leftarrow X} A$, the product over all arrows, i.e., over all morphisms. For this F, $C^*(\mathcal{C}, F)$ is

$$\prod_{Y \leftarrow X_0} A \overset{\sigma_0}{\underset{\sigma_1}{\rightrightarrows}} \prod_{Y \leftarrow X_0 \leftarrow X_1} A \rightrightarrows \cdots ,$$

and we again have the extra degeneracy

$$(\sigma_{-1} f)(Y \longleftarrow X_0 \longleftarrow \cdots \longleftarrow X_p) = f(Y \overset{\text{id}}{\longleftarrow} Y \longleftarrow \cdots \longleftarrow X_p),$$

so the complex is acyclic.

Thus we have constructed functors for H_* and H^* which kill homology in that

Theorem *For all $* > 0$, we have*

$$H_* \left(\mathcal{C}, X \longmapsto \coprod_{X \leftarrow Y} A \right) = 0,$$

$$H^* \left(\mathcal{C}, X \longmapsto \prod_{Y \leftarrow X} A \right) = 0.$$

Example 4.1 Let $\mathcal{C} = \widetilde{G}$ be the category associated to the group G. Consider

$$X \longmapsto \prod_{Y \leftarrow X} A$$

i.e., the image is $\prod_{g \in G} A$, so this functor is the G-module $\mathrm{Maps}(G, A)$, where G acts on the right of G to give a left action on $\mathrm{Maps}(G, A)$.

Example 4.2

$$X \longmapsto \coprod_{X \leftarrow Y} A = \coprod_g A = \mathbb{Z}[G] \otimes_{\mathbb{Z}} A ,$$

where we multiply on the left in $\mathbb{Z}[G]$ for the module structure.

Corollary

$$H^*(G, \text{Maps}(G, A)) = 0 \quad \text{for all } * > 0,$$

$$H_*(G, \mathbb{Z}[G] \otimes_{\mathbb{Z}} A) = 0 \quad \text{for all } * > 0.$$

Note that this corollary plus long exact sequences allow us to compute in practice.

Chapter 5
Effaceable δ-Functors

Suppose $0 \to M' \to M \to M'' \to 0$ is an exact sequence of G-modules. Then we get an exact sequence of complexes of cochains on G

$$0 \longrightarrow C^*(G, M') \longrightarrow C^*(G, M) \longrightarrow C^*(G, M'') \longrightarrow 0$$

with values in these modules and hence get the usual long exact sequence in cohomology

$$0 \longrightarrow H^0(G, M') \longrightarrow H^0(G, M) \longrightarrow H^0(G, M'') \xrightarrow{\delta} H^1(G, M') \longrightarrow \cdots$$

which is natural

$$
\begin{array}{ccccccccc}
0 & \longrightarrow & H^0 M' & \longrightarrow & H^0 M & \longrightarrow & H^0 M'' & \longrightarrow & H^1 M' & \longrightarrow & \cdots \\
& & \downarrow & & \downarrow & & \downarrow & & \downarrow & & \\
0 & \longrightarrow & H^0 N' & \longrightarrow & H^0 N & \longrightarrow & H^0 N'' & \longrightarrow & H^1 N' & \longrightarrow & \cdots
\end{array}
$$

with respect to maps of exact sequences in the usual sense. Last time we saw that if $A \in \mathrm{Ab}$, then the G-module $\mathrm{Maps}(G, A)$, that is, set maps, with the action $(gf)(x) = f(xg)$, had the property that $H^*(G, \mathrm{Maps}(G, A)) = 0$, for all $* > 0$.

A functor $T : \mathcal{C} \to \mathrm{Ab}$ is *effaceable* if for any object X in $\mathcal{C}$ there exists an injection $X \hookrightarrow Y$ in $\mathcal{C}$ so that $T(Y) = 0$. Say a cohomology theory itself is *effaceable* if it is so in positive dimensions.

Proposition $H^*(G, \cdot) : \mathrm{Mod}_G \to \mathrm{Ab}$ *is effaceable, for all* $* > 0$, *where* Mod_G *denotes the category of G-modules.*

© Springer Nature Switzerland AG 2020
R. Penner, *Topology and K-Theory*, Lecture Notes in Mathematics 2262,
https://doi.org/10.1007/978-3-030-43996-5_5

Proof Consider

$$M \longrightarrow \text{Maps}_{\text{set}}(G, M)$$
$$m \longmapsto (x \longmapsto xm)$$

and check this is a G-module map; it is clearly an embedding.   □

Example of Free Groups

Let G be the free group on $\{g_i : i \in I\}$. Then a G-module is the same thing as an abelian group M with a family of automorphisms indexed by I. Recall that $H^0(G, M) = M^G$ and

$$H^2(G, M) = \text{isomorphism classes of extension of } G \text{ by } M.$$

Lift each $g_i \in G$ back to $\tilde{g}_i \in E$. Since G is free, we get a section s to p

$$* \longrightarrow M \longrightarrow E \underset{s}{\overset{p}{\rightleftarrows}} G \longrightarrow *,$$

and s is a group homomorphism. This says the extension splits, and E is the semi-direct product of G acting on M.

Note that in general split extensions of G by M correspond to the zero element in $H^2(G, M)$. The reason is that $s(g_1) s(g_2) = i(f(g_1, g_2)) s(g_1 g_2)$ where f is the two-cocycle belonging to the extension, and $f = 0$ since s is a group map.

Thus, $H^2(G, M) = 0$, for all M if G is free.

Claim $H^*(G, M) = 0$, for all M and all $* \geq 2$ if G is free.

Proof Given M, embed M in N where $H^*(G, N) = 0$ for all $* > 0$ giving

$$0 \longrightarrow M \longrightarrow N \longrightarrow M_1 \longrightarrow 0.$$

Take the long exact sequence. The proof follows from dimension shift and induction.
                                                                                    □

Recall that $H^1(G, M) = Z^1(G, M)/B^1(G, M)$ where

$$Z^1(G, M) = \text{Der}(G, M)$$
$$= \text{derivations } G \to M.$$

Claim $\text{Der}(G, M) = \text{Maps}(I, M)$.

Proof A derivation is determined by the Dg_i, and these can be assigned arbitrarily. To get $D(g_1^{-1})$, note that $D(e) = 0$ so $0 = Dg_1^{-1} + g_1^{-1} Dg_1$ so that $Dg_1^{-1} = -g_1^{-1} D(g_1)$.   □

The exact sequence for cohomology of free groups gives

$$\begin{array}{c} m \longmapsto (i \longmapsto g_i\, m - m) \\ 0 \longrightarrow \underset{\substack{\| \\ M^G}}{H^0(G, M)} \longrightarrow M \longrightarrow \underset{\substack{\| \\ \mathrm{Der}\,(G,M)}}{\mathrm{Maps}\,(I, M)} \longrightarrow H^1(G, M) \longrightarrow 0. \end{array}$$

For instance if $M = \mathbb{Z}$ with trivial G-action, then

$$H^0(G, \mathbb{Z}) = \mathbb{Z} \text{ and } H^1(G, \mathbb{Z}) \cong \mathbb{Z}^n$$

where n is the number of generators of G.

This concludes our discussion for the example of free groups.

Definition A δ-*functor* on the category of G-modules to abelian groups is a collection of so-called additive functors T^q, for $q \geq 0$, together with, for any sequence $0 \to M' \to M \to M'' \to 0$ of G-modules, an assignment of *connecting homomorphisms* $\delta : T^q M'' \to T^{q+1} M'$ so that

(i) $T^0 M' \to T^0 M \to T^0 M'' \overset{\delta}{\to} T^1(M') \to \cdots$ is a complex,
(ii) naturality of connecting homomorphisms with respect to maps of exact sequences.

As an exercise, generalize this definition so that the domain of the functor is any abelian category.

In this definition, an *additive functor* means

$$\mathrm{Hom}_{G\text{-mod}}(M, N) \longrightarrow \mathrm{Hom}_{\mathrm{Ab}}(T^q M, T^q N)$$
$$u \longmapsto T^q(u)$$

is a homomorphism of abelian groups.

Example 5.1 $N \longmapsto \mathrm{Hom}_{G\text{-mod}}(M, N)$ is additive.

Example 5.2 $N \longmapsto N \otimes_{\mathbb{Z}[G]} M$ is additive.

Example 5.3 $N \longmapsto M \otimes_G N$ is *not* additive.

Theorem *Suppose* $\{T^q\}$ *and* $\{U^q\}$ *are* δ-*functors from* G-modules to Ab. *Assume* T^q *is effaceable for all* $q > 0$ *and* $\{T^q\}$ *is* exact, *i.e., (i) in the definition above is in fact exact. Then any natural transformation* $\theta : T^0 \to U^0$ *extends uniquely to a natural transformation of* δ-*functors.*

We shall prove this theorem in the next Chapter.

Corollary *Any two exact effaceable functors that are equal in dimension* 0 *agree everywhere.*

Proof of Corollary Given M, find $0 \to M \to N \to M_1 \to 0$ so that $T^*N = 0$ for all $* > 0$. Then

$$
\begin{array}{ccccccccc}
T^0M & \longrightarrow & T^0N & \longrightarrow & T^0M' & \overset{\delta}{\longrightarrow} & T^1M & \longrightarrow & T^1/N = 0 \\
\downarrow{\scriptstyle\theta} & & \downarrow{\scriptstyle\theta} & & \downarrow{\scriptstyle\theta} & & \Big\downarrow {\scriptstyle\text{exists uniquely}} & & \\
U^0M & \longrightarrow & U^0N & \longrightarrow & U^0M' & \underset{\delta}{\longrightarrow} & U^1M & \longrightarrow & U^1N
\end{array}
$$

Now induct. It remains as an exercise to check functoriality and independence of choice of N. $\square$

Chapter 6
(Co)homology of Cyclic Groups

Homology of Cyclic Groups

Suppose $G = \mathbb{Z}/n\,\mathbb{Z}$, so $\mathbb{Z}[G] = \mathbb{Z}[T]/(T^n - 1)$, T some indeterminate. This holds since $\mathrm{Hom}_{\mathrm{ring}}(\mathbb{Z}[G], A) = \mathrm{Hom}_{\mathrm{groups}}(G, A^{\bullet})$ where $A^{\bullet} =$ groups of units of A and for $G = \mathbb{Z}/n\,\mathbb{Z}$

$$A^{\bullet} = \{a \in A : a^n = 1\}$$
$$= \mathrm{Hom}_{\mathrm{rings}}(\mathbb{Z}[T]/(T^n - 1), A).$$

Now, $T^n - 1 = (T - 1)(T^{n-1} + \ldots + 1)$, so we get two exact sequences of groups and the following identifications:

$(*_1)$

$$0 \longrightarrow (T-1)/(T^n-1) \xrightarrow{\text{mult by } T-1} \mathbb{Z}[T]/(T^n-1) \longrightarrow \mathbb{Z}[T]/(T-1) \longrightarrow 0$$

$$\mathbb{Z}[T]/(T^{n-1} + \ldots + 1) \qquad\qquad\qquad\qquad\qquad \mathbb{Z}$$

where the isomorphism $\mathbb{Z}[T]/(T^{n-1} + \ldots + 1) \to (T-1)/(T^n-1)$ is given by $1 \mapsto T - 1$;

$(*_2)$

$$0 \longrightarrow (T^{n-1} + \ldots + 1)/(T^n - 1) \xrightarrow[\;(T^{n-1}+\ldots+1)\;]{\text{mult by}} \mathbb{Z}[T]/(T^n - 1) \longrightarrow \mathbb{Z}[T]/(T^{n-1} + \ldots + 1) \longrightarrow 0$$

$$\mathbb{Z}[T]/(T-1)$$

$$\mathbb{Z}$$

© Springer Nature Switzerland AG 2020

R. Penner, *Topology and K-Theory*, Lecture Notes in Mathematics 2262,
https://doi.org/10.1007/978-3-030-43996-5_6

where the isomorphism $\mathbb{Z}[T]/(T-1) \to (T^{n-1} + \ldots + 1)/(T^n - 1)$ is given by $1 \mapsto (T^{n-1} + \ldots + 1)$. Note that both these sequences split **over** $\mathbb{Z}$ since the terms are free. Note also that $(*_2)$ is a "resolution" of the leftmost term in $(*_1)$.
Define

$$I(G) = \mathbb{Z}[T]/(T^{n-1} + \ldots + 1)$$
$$= \text{augmentation ideal for } G = \mathbb{Z}/n\,\mathbb{Z},$$
$$\mathbb{Z}(G) = \mathbb{Z}[T]/(T^n - 1)$$
$$= \mathbb{Z}[G] \text{ for } G = \mathbb{Z}/n\,\mathbb{Z}, \text{ as above.}$$

The sequences $(*_1)$ and $(*_2)$ become

$$(**_1) \qquad\qquad 0 \longrightarrow I(G) \longrightarrow \mathbb{Z}(G) \longrightarrow \mathbb{Z} \longrightarrow 0\,,$$

$$(**_2) \qquad\qquad 0 \longrightarrow \mathbb{Z} \longrightarrow \mathbb{Z}(G) \longrightarrow I(G) \longrightarrow 0\,.$$

Now suppose M is a G-module, $G = \mathbb{Z}/n\,\mathbb{Z}$, and tensor $(**_1)$ and $(**_2)$ with M to get two exact sequences of G-modules

$(\dagger_1)$

$$0 \longrightarrow I(G) \otimes_{\mathbb{Z}} M \xrightarrow{(T-1)\otimes m \,\mapsto\, 1\otimes m} \mathbb{Z}[G] \otimes_{\mathbb{Z}} M \xrightarrow{1\otimes m \,\mapsto\, m} M \longrightarrow 0$$

$(\dagger_2)$

$$0 \longrightarrow M \xrightarrow{m \,\mapsto\, \sum_{i=0}^{n-1} T^i \otimes m} \mathbb{Z}[G] \otimes_{\mathbb{Z}} M \xrightarrow{1\otimes m \,\mapsto\, (T-1)\otimes m} I(G) \otimes_{\mathbb{Z}} M \longrightarrow 0$$

where the G-action on M is as usual and the G-module structure of $A \otimes_{\mathbb{Z}} B$ is $g(a \otimes b) = ga \otimes gb$; check these are G-module sequences.

Moreover, if M is a G-module, then $\mathbb{Z}[G] \otimes_{\mathbb{Z}} M$ with action $x(g \otimes m) = xg \otimes m$ is isomorphic to $\mathbb{Z}[G] \otimes_{\mathbb{Z}} M$ with action $x(g \otimes m) = xg \otimes xm$, where $g \otimes m \mapsto g \otimes gm$ and $g \otimes g^{-1}m' \leftarrow g \otimes m'$. Recall we showed $H_*(G, \mathbb{Z}[G] \otimes_{\mathbb{Z}} A) = 0$, for all $* > 0$. We use this and long exact sequences to compute the homology of cyclic groups. Long exact sequences of $(\dagger_1)$ and $(\dagger_2)$ respectively give

$$H_q(G, \mathbb{Z}[G] \otimes M) \longrightarrow H_q(G, M) \xrightarrow{\partial} H_{q-1}(G, I[G] \otimes M) \longrightarrow H_{q-1}(G, \mathbb{Z}[G] \otimes M)$$

so the first ∂ is an isomorphism for $q \geq 2$, and

$$H_q(G, \mathbb{Z}[G] \otimes M) \longrightarrow H_q(G, I(G) \otimes M) \xrightarrow{\partial} H_{q-1}(G, M) \longrightarrow H_{q-1}(G, \mathbb{Z}[G] \otimes M)$$

so the second ∂ is an isomorphism for $q \geq 2$.

We thus get canonical isomorphisms

$$\boxed{H_q(G, M) \longrightarrow H_{q-2}(G, M), \qquad \text{for all } q > 2}$$

and we finally compute $H_q(G, M)$, for $q = 0, 1, 2$.

Exercise

$$\boxed{H_0(G, M) = \mathbb{Z} \bigotimes_{\mathbb{Z}[G]} M = M/\{(g-1)m\}}$$

i.e., the largest quotient on which the group acts trivially. Indeed,

$$C_1(G, M) \rightrightarrows C_0(G, M)$$

$$\coprod_{* \xleftarrow{g} *} M \overset{d_0}{\underset{d_1}{\rightrightarrows}} M$$

and

$$d_0 : g \otimes m \longmapsto m,$$
$$d_1 : g \otimes m \longmapsto gm.$$

Thus,

$$H_0(G, \mathbb{Z}[G] \otimes M) = \mathbb{Z} \bigotimes_{\mathbb{Z}[G]} (\mathbb{Z}[G] \otimes_{\mathbb{Z}} M) = M.$$

Now, we have

$$0 \longrightarrow H_1(G, I \otimes M) \overset{\partial}{\longrightarrow} H_0(G, M) \longrightarrow H_0(G, \mathbb{Z}[G] \otimes M) \longrightarrow H_0(G, I \otimes M) \longrightarrow 0$$

$$M/\{(g-1)m\}$$

$$M/(T-1)M \overset{n-1}{\underset{\text{mult by} \sum_{i=0} T^i}{\longrightarrow}} M$$

The map given by multiplication by $\sum_{i=0}^{n-1} T^i$ is called the *norm*: $M \to M$.

Thus $H_0(G, I \otimes M) = M/\text{Im}\{\text{norm} : M \to M\}$ and

$$H_1(G, I \otimes M) = \frac{\text{Ker}\{\text{norm} : M \to M\}}{\text{Im}\{(T-1) : M \to M\}}.$$

Since $H_2(G, M) = H_1(G, I \otimes M)$, we get

$$\boxed{H_2(G, M) = \frac{\text{Ker}\{\text{norm} : M \to M\}}{\text{Im}\{(T-1) : M \to M\}}}$$

Exercise
$$\boxed{H_1(G, M) = \frac{\text{Ker}\{(T-1) : M \to M\}}{\text{Im}\{\text{norm} : M \to M\}}}$$

Cohomology of Cyclic Groups
Recall $H^*(G, \text{Maps}\,(G, A)) = 0$, for all $* > 0$, and for G finite abelian

$$\text{Maps}\,(G, M) = \text{Hom}_{\mathbb{Z}}(\mathbb{Z}[G], M) = \text{Hom}_{\mathbb{Z}}(\mathbb{Z}[G], \mathbb{Z}) \otimes_{\mathbb{Z}} M.$$

Thus, from $(**_1)$ and $(**_2)$, which we recall are split, we get two exact sequences

$$\begin{array}{c}
M \\
\Vert \\
0 \longrightarrow \text{Hom}_{\mathbb{Z}}(\mathbb{Z}, M) \longrightarrow \text{Hom}_{\mathbb{Z}}(\mathbb{Z}[G], M) \longrightarrow \text{Hom}_{\mathbb{Z}}(I, M) \longrightarrow 0
\end{array}$$

$$\begin{array}{c}
0 \longrightarrow \text{Hom}_{\mathbb{Z}}(I, M) \longrightarrow \text{Hom}_{\mathbb{Z}}(\mathbb{Z}[G], M) \longrightarrow \text{Hom}_{\mathbb{Z}}(\mathbb{Z}, M) \longrightarrow 0 \\
\Vert \\
M
\end{array}$$

and as above $\text{Hom}_{\mathbb{Z}}(\mathbb{Z}[G], M)$ has trivial higher cohomology.

Exercise (*if interested*)

$$H^q(G, M) \xrightarrow{\cong} H^{q+2}(G, M), \qquad \text{for } q \geq 1,$$

$$H^2(G, M) = \frac{\text{Ker}\,\{(T-1) : M \to M\}}{\text{Im}\,\{\text{norm} : M \to M\}},$$

$$H^1(G, M) = \frac{\text{Ker}\,\{\text{norm} : M \to M\}}{\text{Im}\,\{(T-1) : M \to M\}},$$

and we know $H^0(G, M) = M^G$, as before, completing this discussion of cyclic groups.

Recall the

Theorem *Suppose* $\{T^q, q \geq 0\}$ *is an exact δ-functor and* T^q *is effaceable for all* $q > 0$ *and* $\{U^q, q \geq 0\}$ *is any δ-functor. Then any natural transformation* $\theta^0 :$ $T^0 \to U^0$ *extends uniquely to a natural transformation of δ-functors.*

Proof It suffices to extend to $q = 1$ and check properties by induction.

1: Uniqueness: Suppose given M and find exact

$$0 \longrightarrow M \longrightarrow M' \longrightarrow M'' \longrightarrow 0 \quad \text{so that} \quad T^1 M' = 0.$$

Then we have

$$
\begin{array}{ccccccc}
T^0 M' & \longrightarrow & T^0 M'' & \overset{\delta}{\longrightarrow} & T^1 M & \longrightarrow & T^1 M' = 0 \\
\downarrow{\scriptstyle \theta^0} & & \downarrow{\scriptstyle \theta^0} & & \downarrow{\scriptstyle \theta^1} & & \\
U^0 M' & \longrightarrow & U^0 M'' & \overset{\delta}{\longrightarrow} & U^1 M & &
\end{array}
$$

with the top row exact. A diagram chase gives uniqueness of θ^1.

Defining θ^1 as before in the obvious way, we must check:

Claim 6.2 Well-defined.

Claim 6.3 Natural transformation.

Claim 6.4 Compatible with δ.

Check 6.2 We have the

Claim Given $M \hookrightarrow M'$ and $M \hookrightarrow M_1'$ with $T^1 M' = T^1 M_1' = 0$, then we can embed these maps in a diagram

$$
\begin{array}{ccc}
M & \hookrightarrow & M' \\
\downarrow & & \downarrow \\
M_1' & \hookrightarrow & M_2'
\end{array}
$$

with $T^1 M_2' = 0$.

Proof Take the push out, i.e., take $\widehat{M_2'} = M' \oplus M_1'/(\operatorname{Im} M \oplus -\operatorname{Im} M)$ and embed it $\widehat{M_2'} \hookrightarrow M_2'$ by effaceability. $\square$

Thus to show $\theta^1 : T^1 M \to U^1 M$ is the same for $M \hookrightarrow M'$ and $M \hookrightarrow M'_1$, we may assume

commutes. Thus we have

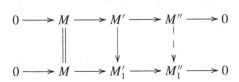

which gives the following cube of arrows

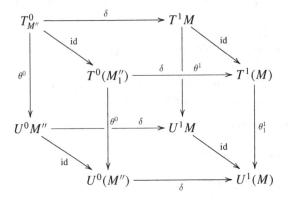

and this all commutes by definition except perhaps the rightmost face, which thus also commutes, as desired.

Check 6.3 Suppose we have

$$0 \longrightarrow M \longrightarrow M' \longrightarrow M'' \longrightarrow 0$$
$$\downarrow{\scriptstyle u} \qquad \downarrow \qquad \downarrow$$
$$0 \longrightarrow M_1 \longrightarrow M'_1 \longrightarrow M''_1 \longrightarrow 0$$

We get the same cube as above with identity maps replaced by u_* and all appearances of M in parentheses having subscript "1".

Check 6.4 We must check that an exact

$$0 \longrightarrow M \longrightarrow N \longrightarrow Q \longrightarrow 0$$

gives a commutative

$$T^0 Q \xrightarrow{\delta} T^1 M$$

$$\downarrow \theta^0 \qquad\qquad \downarrow \theta^1$$

$$U^0 Q \xrightarrow{\delta} U^1 M$$

To this end we resolve

so that $T^1 M' = 0$.

Exercise Finish the argument.   □

Chapter 7
An Application to the Schur–Zassenhaus Theorem

We have the

Theorem *Given* $\{T^q\}$ *an exact δ-functor with* T^q *effaceable, for* $q > 0$, *and* $\{U^q\}$ *any δ-functor. Then any natural transformation* $\theta^0 : T^0 \to U^0$ *extends uniquely to a natural transformation of δ-functors.*

Example 7.1 Suppose $f : H \to G$ is a group homomorphism. Then any G-module M is also an H-module, denoted f^*M. We define two δ-functors

$$T^q(M) = H^q(G, M)$$
$$U^q(M) = H^q(H, M) \text{ with } M \text{ regarded as } f^*M.$$

from the category of G-modules to the category of abelian groups. Automatically a short exact sequence of G-modules is a short exact sequence of H-modules. Then

$$H^0(\underset{\substack{\| \\ M^G}}{G}, M) \subset H^0(\underset{\substack{\| \\ M^H}}{H}, M)$$

is a natural transformation of δ-functors. By the theorem, there exists a unique morphism of δ-functors

$$\mathrm{Res}_{f:H \to G} : H^q(G, M) \longrightarrow H^q(H, M)$$

extending from G to H.

Formulae Given $K \xrightarrow{u} H \xrightarrow{f} G$, we have

$$\mathrm{Res}_{K \to M} \, \mathrm{Res}_{H \to G} = \mathrm{Res}_{K \to G}$$

by the uniqueness assertion, since it is true in degree 0. We could define the restriction on the cochain level, and this must agree, again by uniqueness.

Example 7.2 Transfer Map Let $H < G$ be a subgroup of finite index and let M be a G-module. We have the *transfer*

$$H^0(\underset{\substack{\| \\ M^H}}{H}, M) \longrightarrow H^0(\underset{\substack{\| \\ M^G}}{G}, M),$$

© Springer Nature Switzerland AG 2020
R. Penner, *Topology and K-Theory*, Lecture Notes in Mathematics 2262,
https://doi.org/10.1007/978-3-030-43996-5_7

defined as follows. If $m \in M^H$ and $g \in G$, then gm depends only on the coset gH of g, where $(gh)m = g(hm) = gm$. Thus if $g_1, \ldots, g_n$ are coset representatives, so $G = \coprod_{i=1}^{n} g_i\, H$, then the element

$$\sum_{i=1}^{n} g_i\, m$$

is independent of the choice of coset representatives. Write it as $\sum_{gH\in G/H} (gH)\, m$. This element is invariant under all $g \in G$. In this way we get a homomorphism

$$M^H \longrightarrow M^G$$

called the *transfer*, denoted variously by

$$\mathrm{Ver}_{H\to G}\,,\ \ \mathrm{Ind}_{H\to G}\,.$$

Check the hypotheses of the theorem so as to extend to higher dimensions. Exactness is clear, and the only point that remains to be checked is that $H^q(H, \cdot)$ from the category of G-**modules** is effaceable. As before, we have

$$M \longrightarrow \mathrm{Maps}\,(G, M)\,,$$

and it is enough to show that for any abelian group A that as an H-module

$$\mathrm{Maps}\,(G, A) = \prod_{gH\in G/H} \mathrm{Maps}\,(gH, A)$$

$$\| \wr$$

$$\prod_{gH\in G/H} \mathrm{Maps}\,(H, A)$$

and Maps (H, A) has trivial $H^+(H, \mathrm{Maps}(H, A))$, as proved at the end of Chapter 4. We thus obtain $\mathrm{Ver}_{H\to G} : H^*(H, M) \to H^*(G, M)$.

Formulae

Exercise 7.1 Transitivity If $K < H < G$ with

$$[H : K] < \infty \quad [H : G] < \infty\,,$$

then

$$\mathrm{Ver}_{H\hookrightarrow G}\, \mathrm{Ver}_{K\hookrightarrow H} = \mathrm{Ver}_{K\hookrightarrow G}\,.$$

Exercise 7.2 If $H < G$ is finite index, then

$$H^*(G, M) \xrightarrow{\text{Res}_{H \hookrightarrow G}} H^*(H, M) \xrightarrow{\text{Ver}_{H \hookrightarrow G}} H^*(G, M)$$

is multiplication by $[G : H]$.

Exercise 7.3 Mackey Formula Suppose

$$
\begin{array}{ccc}
ghg^{-1} & \longmapsto & h \\
K \cap gHg^{-1} & \longrightarrow & H \\
\cap & & \cap \text{ finite index} \\
K & \underset{\text{inclusion}}{\longrightarrow} & G
\end{array}
$$

Then

$$\text{Res}_{K \to G} \text{Ver}_{H \hookrightarrow G} = \sum_{KgH \in K\backslash G/H} \text{Ver}_{K \cap gHg^{-1} \to K} \text{Res} \qquad \begin{array}{c} K \cap gHg^{-1} \to H \\ x \mapsto g^{-1}xg \end{array} .$$

This is not fantastically useful.

Interesting special case: $H = K \lhd G$ a normal subgroup of finite index, then

$$\text{Res}_{H \hookrightarrow G} \text{Ver}_{H \hookrightarrow G} = \sum_{gH \in G/H} \text{Res} \qquad \begin{array}{c} H \to H \\ h \mapsto g^{-1}hg \end{array} .$$

Proof of 2 Check in degree zero and use the Theorem:

$$
\begin{array}{ccccl}
M^G & \xrightarrow{\text{Res}} & M^H & \xrightarrow{\text{Ver}} & M^G \\
m & \longmapsto & m & \longmapsto & \sum_{gH \in G/H} (gH)\, m = [G : H]\, m \quad \text{for } m \in M^G.
\end{array}
$$

$\square$

Corollary of 2 *Suppose that G is a finite group and M is a G-module so that multiplication by $|G|$ is an isomorphism of M. Then $H^+(G, M) = 0$, where H^+ means H^* for $* > 0$.*

Proof Take $H = \{e\}$. Then we have the commutative

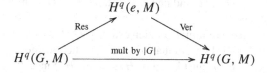

$$
\begin{array}{ccc}
 & H^q(e, M) & \\
\text{Res} \nearrow & & \searrow \text{Ver} \\
H^q(G, M) & \xrightarrow{\text{mult by } |G|} & H^q(G, M)
\end{array}
$$

Multiplication by $|G|$ is an isomorphism and $H^*(e, M) = 0$, for $q > 0$, as the trivial group e is hilariously free. $\square$

Theorem (Schur–Zassenhaus) *Any extension* $* \longrightarrow H \longrightarrow G \xrightarrow{p} Q \longrightarrow *$ *of finite groups, where* $|H|$ *and* $|Q|$ *are relatively prime, splits, i.e., there exists a homomorphism* $s : Q \to G$ *so that* $ps = $ id.

In particular, if $H \triangleleft G$, for G finite and $|H|$ relatively prime to $[G : H]$, then there exists a subgroup complementary to H.

Proof (*in the spirit of finite group-theory*) Let G be a minimal counter-example. Let P be a Sylow p-subgroup of G for a prime p dividing $|H|$. Then since $|H|$ and $|Q|$ are relatively prime, we must have $P \subset H$.

Claim $N_G(P)$ maps onto Q, for, given $q \in Q$, lift it to g; compare gPg^{-1} and P, both Sylow p-subgroups of H, so they are conjugate in H and there exists $h \in H$ with $gPg^{-1} = hPh^{-1}$ which implies $h^{-1}gP(h^{-1}g)^{-1} = P$, so $h^{-1}g \in N_G H$, the normalizer of H in G, and $h^{-1}g \xmapsto{p} q$. Thus we have

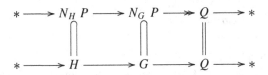

where the kernel of the top map is $N_H P$.

By minimality, if $N_G(P) < G$, then $N_G(P)$ would contain a subgroup mapping onto Q, a contradiction since the top sequence would then split, whence the bottom as well. We conclude that

$$N_G P = G .$$

Transitivity argument Suppose we can find a subgroup K so that $1 < K < H$ with $K \triangleleft G$. Then

$$* \longrightarrow H/K \longrightarrow G/K \longrightarrow Q \longrightarrow * .$$

By minimality, this splits. Let $A \subset G/K$ map isomorphically onto Q. Then $A = B/K$ where $B \subsetneqq G$ with $B \twoheadrightarrow Q$. It is therefore an extension of Q by $H \cap B$, which is of order prime to Q. By minimality again, this extension

$$* \longrightarrow B \cap H \longrightarrow B \twoheadrightarrow Q \longrightarrow *$$

splits, giving a section $Q \to B$ and hence a section $Q \to G$, contradiction. Thus K as above does not exist.

We conclude that for a minimal counter-example G, H has no proper subgroups K which are normal in G. In particular, P, which is the **unique** Sylow p-subgroup of G since it is normal in G must be all of H, so $P = H$. Since P has no characteristic subgroups, i.e., invariant under all automorphisms of the group, it must be an elementary **abelian** p-group.

Now, use that $H^2(Q, H)$ classifies the extensions for H abelian. H is a p-group and p is relatively prime to $|Q|$, so the Corollary of 2 above implies that $H^+(Q, H) = 0$. A minimal counter-example therefore cannot exist. $\square$

Chapter 8
The Yoneda Lemma

Let $\mathcal{C}$ be some category and fix some $X \in \mathrm{Ob}\,\mathcal{C}$. Define $\mathcal{C}^{\mathrm{op}} = opposed\ category$, which has the same objects but with $\mathrm{Hom}_{\mathcal{C}^{\mathrm{op}}}(X, Y) = \mathrm{Hom}_{\mathcal{C}}(Y, X)$.

Define $h_X : \mathcal{C}^{\mathrm{op}} \to$ Sets, i.e., just a contravariant functor $\mathcal{C} \to$ Sets, given by

$$h_X : Y \longmapsto \mathrm{Hom}_{\mathcal{C}}(Y, X).$$

Define

$$h^X : \mathcal{C} \longrightarrow \text{Sets}$$
$$Y \longmapsto \mathrm{Hom}_{\mathcal{C}}(X, Y).$$

Yoneda Lemma If $F : \mathcal{C}^{\mathrm{op}} \to$ Sets is a functor, then the set $\mathrm{NatTrans}(h_X, F)$ of natural transformations from h_X to F is in one-to-one correspondence with $F(X)$ given by

$$(\theta : h_X \to F) \longmapsto \theta(X)(\mathrm{id}_X)$$
$$\theta \text{ so that } \theta(Y)(f) = F(f)\xi \longleftarrow\!\shortmid\ \xi.$$

In a diagram,

$$
\begin{array}{ccc}
\mathrm{Hom}(Y, X) = h_X(Y) & \xrightarrow{\ \theta(Y)\ } & F(Y) \\
\uparrow{\scriptstyle f_*} & & \uparrow{\scriptstyle F(f)} \\
\mathrm{Hom}(X, X) = h_X(X) & \xrightarrow{\ \theta(X)\ } & F(X)
\end{array}
$$

where

$$\theta(Y)(f) = F(f)\,\theta(X)\,(\mathrm{id}_X).$$

© Springer Nature Switzerland AG 2020

R. Penner, *Topology and K-Theory*, Lecture Notes in Mathematics 2262,

https://doi.org/10.1007/978-3-030-43996-5_8

Corollary NatTrans(h_X, h_Y) *is in one-to-one correspondence with* Hom(X, Y).
Modulo set-theory, we have a functor as follows

$$\mathcal{C} \longrightarrow \text{Funct}(\mathcal{C}^{\text{op}}, \text{Sets})$$
$$X \longmapsto h_X,$$

*where Funct as a category has its morphisms given by natural transformations, and
the corollary says*

$$\text{Hom}_{\text{Funct}(\mathcal{C}^{\text{op}}, \text{Sets})}(h_X, h_Y) = \text{Hom}_{\mathcal{C}}(X, Y).$$

Thus $X \mapsto h_X$ gives an embedding of $\mathcal{C}$ in Funct($\mathcal{C}^{\text{op}}$, Sets).

Definition A functor $F : \mathcal{C}^{\text{op}} \to$ Sets is *representable* if it is isomorphic to a functor
of the form h_X. According to Yoneda's Lemma, an isomorphism of $h_X \simeq F$ is given
by some $\xi \in F(X)$. Thus F is representable when there exist $X \in \text{Ob}\,\mathcal{C}$ and $\xi \in F(X)$
so that for all $Y \in \text{Ob}\,\mathcal{C}$,

$$\text{Hom}_{\mathcal{C}}(Y, X) \xrightarrow{\simeq} F(Y)$$
$$f \longmapsto F(f)\xi.$$

This says (X, ξ) have the following universal property: for any $Y \in \text{Ob}\,\mathcal{C}$ and $\eta \in$
$F(Y)$ there is a unique map $f : Y \to X$ so that η is induced from ξ.

General exercise Re-write universal constructions in terms of representable functors.

Example 8.1 A *final* (or *terminal*) object of $\mathcal{C}$, call it e, is an object so that for all
$Y \in \mathcal{C}$ there exists a unique $Y \to e$, and e represents the functor $Y \to *$.

Example 8.2 The product $\Pi_I X_i \in \text{Ob}\,\mathcal{C}$ of $X_i \in \text{Ob}\,\mathcal{C}$, for $i \in I$, equipped with
maps $\text{pr}_i : \Pi_I X_i \to X_i, i \in I$, has the universal property that for all Y and $f_i : Y \to$
X_i, for all $i \in I$, there exists a unique map $Y \xrightarrow{f} \Pi X_i$ so that $\text{pr}_i f = f_i$. The
functor represented is $\prod_{i \in I} h_{X_i}$.

Example 8.3 Let I be a small category and $i \mapsto X_i$ a functor from I to $\mathcal{C}$. Then
$\varprojlim X_i$ denotes an object in $\mathcal{C}$ equipped with maps

$$p_i : \varprojlim X_i \longrightarrow X_i$$

compatible with the morphisms in I in the sense that for all $\alpha : i \to i'$ in I, the
diagram

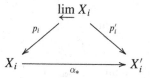

commutes and enjoying the obvious universal property. The "new" terminology for this is simply $\varprojlim = \lim$ and $\varinjlim = \operatorname{colim}$. The functor being so represented assigns to Y all

$$(f_i) \in \prod_{i \in \operatorname{Ob} I} \operatorname{Hom}(Y, X_i)$$

so that for all $\alpha : i \mapsto i'$ we have $\alpha_* f_i = f_{i'}$. We can write this as

$$\operatorname{Ker} \left\{ \prod_{i \in \operatorname{Ob} \mathcal{C}} \operatorname{Hom}(Y, X_i) \rightrightarrows \prod_{i' \xleftarrow{\alpha} i} \operatorname{Hom}(Y, X_{i'}) \right\}$$

$$f(i) \begin{array}{c} \longmapsto \\ \longmapsto \end{array} \begin{array}{c} \left(\left(i' \xleftarrow{\alpha} i \right) \longmapsto f_{i'} \right) \\ \left(\left(i' \xleftarrow{\alpha} i \right) \longmapsto \alpha_* f_i \right) \end{array}$$

where Ker here denotes *equalizer*, i.e., elements having the same image under each map.

If this kernel is written $\varprojlim_I \operatorname{Hom}(Y, X_i)$ then

$$\operatorname{Hom}_{\mathcal{C}} \left(Y, \varprojlim X_i \right) = \varprojlim_I \operatorname{Hom}_{\mathcal{C}}(Y, X_i) .$$

Note that if $F : \mathcal{C} \to \operatorname{Sets}$, then

$$\varprojlim_{\mathcal{C}} F = \operatorname{Ker} \left\{ \prod_{X_0} F(X_0) \rightrightarrows \prod_{X_0 \leftarrow X_1} F(X_0) \right\} ,$$

and this coincides with $H^0(\mathcal{C}, F)$ for F an abelian group functor, i.e., in general

$$H^0(\mathcal{C}, F) = \varprojlim_{\mathcal{C}} F .$$

For example, a G-set is a functor

$$\tilde{G} \longrightarrow \operatorname{Sets}$$
$$* \longmapsto S$$

and then

$$\varprojlim_{\widetilde{G}} S = S^G .$$

In general by Yondea, we have $\mathrm{Hom}_{\mathrm{Funct}\,(\mathcal{C},\mathrm{Sets})}(h^X, F) = F(X)$ and get an embedding

$$\mathcal{C}^{\mathrm{op}} \longrightarrow \mathrm{Funct}(\mathcal{C}, \mathrm{Sets})$$
$$X \longmapsto h^X .$$

An *initial* object ϕ is so that $\mathrm{Hom}(\phi, Y) = *$ for all Y.

For instance:

(1) *Direct sum* $\coprod_I X_i$: for all $i \in I$, we have $X_i \xrightarrow{in_i} \coprod_I X_i$ with the appropriate universal property.

(2) *Direct limit* (or *colimit*): $\varinjlim_I X_i$ together with $Y_i \xrightarrow{in_i} \varinjlim_I X_i$ compatible with the arrows of I and the appropriate universal property.

(1) represents $\mathrm{Hom}\left(\coprod_I X_i, Y\right) = \prod_I \mathrm{Hom}(X_i, Y)$, and

(2) represents $\mathrm{Hom}_{\mathcal{C}}\left(\varinjlim_I X_i, Y\right) = \varprojlim_I \mathrm{Hom}_{\mathcal{C}}(X_i, Y)$.

Example With $\mathcal{C} = \widetilde{G}$, a functor $S : \widetilde{G} \to \mathrm{Sets}$ is the same as a G-set S, and a functor $M : \widetilde{G} \to \mathrm{Ab}$ is a G-module M, where

$$\varprojlim_{\widetilde{G}} S = S^G \quad \text{and} \quad \varprojlim_{\widetilde{G}} M = M^G .$$

But what about

$$\varinjlim_{\widetilde{G}} S = ? \quad \text{and} \quad \varinjlim_{\widetilde{G}} M = ?$$

Now, $\varinjlim_{\widetilde{G}} S$ is a set together with arrows $S \to \varinjlim S$ compatible with the arrows in $\widetilde{G}$, that is,

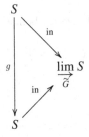

Thus $\varinjlim_{\widetilde{G}} S$ is the set of orbits $G \backslash S$, and

$$\varinjlim_{\widetilde{G}} S = M/\{gm - m : g \in G,\ m \in M\}$$

$$= \text{largest quotient module of } M \text{ on which } G \text{ acts trivially.}$$

Thus, we must specify the category in which we are computing $\varinjlim$, as they do not in general agree.

Adjoint Functors

Suppose C and C' are categories. A *pair of adjoint functors* is $C \underset{G}{\overset{F}{\rightleftarrows}} C'$ together with a **natural** bijection

$$(*) \qquad\qquad \text{Hom}_{C'}(F(X), Y) \xrightarrow{\sim} \text{Hom}_C(X, G(Y)).$$

for each $X \in \text{Ob}\,C$, $Y \in \text{Ob}\,C'$. F is the *left adjoint* and G *right adjoint* of the pair.
 By Yoneda, $(*)$ is given either by natural transformations

$$\alpha : FG \longrightarrow \text{id}_{C'}$$

or

$$\beta : \text{id}_C \longrightarrow GF.$$

Easy to check that

$$F = F\,\text{id}_C \xrightarrow{F \cdot \beta} FGF \xrightarrow{\alpha \cdot F} \text{id}_{C'} F = F$$

and

$$G = \text{id}_C G \xrightarrow{\beta \cdot G} GFG \xrightarrow{G \cdot \alpha} G\,\text{id}_{C'} = G$$

are both the identity. Thus we could alternatively define adjoint functors to be (F, G) together with α and β so that these composites are the identity. To see this given α, β, we have

$$\text{Hom}(FX, Y) \xrightarrow{G} \text{Hom}(GFX, GY) \xrightarrow{\beta^*} \text{Hom}(X, GY) \xrightarrow{F} \text{Hom}(FX, FGY) \xrightarrow{\alpha^*} \text{Hom}(FX, Y)$$

By Yoneda, take $Y = F(X)$. Then

$$\text{id}_X \longmapsto \text{id}_{GFX} \qquad \text{and} \qquad \beta \longmapsto F \cdot \beta \longmapsto \alpha F \cdot \beta,$$

and equivalence of the two definitions follows.

Kan Formulae for Computing Adjoint Functors

Let C and C' be small categories and $f : C \to C'$. Consider

$$\text{Funct}\,(C, \text{Sets}) \xleftarrow{f^*} \text{Funct}\,(C', \text{Sets})$$

i.e.,

$$(f^*F)(X) = F(f(X)).$$

This f^* has two adjoints $f_!$ and f_*.

Theorem f^* *has a left adjoint $f_!$ (pronounced f shriek) and right adjoint f_*.*

We shall discuss this next time.

Chapter 9
Kan Formulae

Suppose $\mathcal{C} \underset{G}{\overset{F}{\rightleftarrows}} \mathcal{C}'$ is a pair of adjoint functors where we are given natural transformations

$$\mathrm{Hom}_{\mathcal{C}'}(FX, Y) \simeq \mathrm{Hom}_{\mathcal{C}}(X, GY).$$

Example 9.1

$$(\text{Sets}) \underset{G=\text{forgetful}}{\overset{F=\text{free group}}{\rightleftarrows}} (\text{Groups})$$

and

$$\mathrm{Hom}_{\text{groups}}(FS, G) = \mathrm{Hom}_{\text{Sets}}(S, G_{\text{forgotten}}).$$

Example 9.2 Suppose $A \xrightarrow{u} B$ is a homomorphism of rings with unit, and Mod_A is the category of (left) A-modules. We have

$$\mathrm{Mod}_A \xleftarrow{u^*} \mathrm{Mod}_B$$

where u^* is restriction of scalars with respect to u, i.e., $M \mapsto M$ and we have $am = u(a)m$.

Then u^* has a left adjoint given by base extension

$$\mathrm{Mod}_A \longrightarrow \mathrm{Mod}_B,$$

i.e., $M \longmapsto B \otimes_A M$ B is a **right** A module via u, and

$$\mathrm{Hom}_B(B \otimes_A M, N) = \mathrm{Hom}_A(M, u^*N).$$

And u^* has also a right adjoint

© Springer Nature Switzerland AG 2020
R. Penner, *Topology and K-Theory*, Lecture Notes in Mathematics 2262,
https://doi.org/10.1007/978-3-030-43996-5_9

$\text{Mod}_A \longrightarrow \text{Mod}_B$

$\quad M \longmapsto \text{Hom}_A(B, M) = \{f : B \to M \mid f(u(a)b) = uf(b) \text{ and } f \text{ is additive}\}$

where $\text{Hom}_A(B, M)$ is a left B-module by

$$bf(b_1) = f(b_1 b), \text{ i.e., use right action on } B,$$

and

$$\text{Hom}_B(N, \text{Hom}_A(B, M)) = \text{Hom}_A(u^* N, M).$$

Exercise Check this. The idea is that the left-hand side is $\text{Hom}_A(B \otimes_B N, M)$ and $B \otimes_B N = N$; see Cartan and Eilenberg.

Summary of Example 9.5:

$$M \longmapsto B \otimes_A M$$

$$\text{Mod}_A \xleftarrow{\quad u^* \quad} \text{Mod}_B$$

$$M \longmapsto \text{Hom}_A(B, M)$$

with canonical maps; recall Yoneda's lemma that canonical maps are given by $FG \to$ id and $GF \leftarrow$ id

$$u^*(B \otimes_A M) \longleftarrow M \qquad B \otimes_A u^* N \longrightarrow N$$
$$1 \otimes m \quad \longleftarrow\!\shortmid\ m \qquad\quad b \otimes n \ \longmapsto\ n$$

and

$$\text{Hom}_A(B, u^* N) \longrightarrow N \qquad u^* \text{Hom}_A(B, N) \longleftarrow N$$
$$\qquad f \qquad\quad \longmapsto f(1) \qquad (b \mapsto bn) \quad \longleftarrow\!\shortmid\ n.$$

Example 9.3 Consider the category of topological spaces and continuous maps. Let I be compact. Then

$$\text{Hom}(X \times I, Y) = \quad \text{Hom}(X, Y^I)$$

with Y^I given the compact-open topology.

Changing gears now, consider the following situation: $\mathcal{C}$ and $\mathcal{C}'$ are small categories and $f : \mathcal{C} \to \mathcal{C}'$ a functor. Set $\widehat{\mathcal{C}} = \text{Funct}(\mathcal{C}, \text{Sets})$, so we have $\widehat{\mathcal{C}} \xleftarrow{f^*} \widehat{\mathcal{C}'}$, $G \circ f \leftarrow\!\shortmid G$.

Kan's Formula[1]

f^* has left adjoint $f_!$ and right adjoint f_* given by

$$(f_! F)(Y) = \varinjlim_{(X,u:fX \to Y)} F(X)$$

$$(f_* F)(Y) \quad \varprojlim_{(X, Y \to fX)} F(X).$$

For $Y \in \mathrm{Ob}\, \mathcal{C}'$, consider the category whose objects are pairs (X, u), $X \in \mathrm{Ob}\, \mathcal{C}$ and $u : fX \to Y$ a morphism in $\mathcal{C}'$ and

$$\mathrm{Hom}((X, u), (X', u')) = \left\{ v : X \to X' \mid \begin{array}{c} fX \\ \downarrow f(v) \quad \searrow^u \\ \qquad\qquad Y \ \text{commutes} \\ \nearrow_{u'} \\ fX' \end{array} \right\}$$

Call this the *left-fiber of* $f : \mathcal{C} \to \mathcal{C}'$ *over* Y. As an exercise, we can check that this is a category.

Similarly, we have the *right-fiber* of $f : \mathcal{C} \to \mathcal{C}'$ over Y consisting of all pairs $(X, u : Y \to fX)$, again a category .

Each of these is different from the *fiber* of $f : \mathcal{C} \to \mathcal{C}'$ over Y, which consists of all $X \in \mathrm{Ob}\, \mathcal{C}$ with $f(X) = Y$, and morphisms sit over the identity of Y. The fiber is denoted simply $f^{-1}Y$.

Note that $f^{-1}Y$ is a subcategory of both the left fiber and the right fiber.

Example 9.4 Suppose $\mathcal{C}$ and $\mathcal{C}'$ are *discrete* categories, i.e., objects are sets and all arrows are identity maps. Then $f : \mathcal{C} \to \mathrm{Sets}$ is a family of sets $F(X)$ indexed by $X \in \mathrm{Ob}\, \mathcal{C}$. So we can think of F as just a set over $\mathrm{Ob}\, \mathcal{C}$, i.e., $\begin{array}{c} F \\ \downarrow \\ {\scriptstyle \mathrm{Ob}\, \mathcal{C}} \end{array}$, so in this case $\widehat{\mathcal{C}}$ is

the category of sets over $\mathrm{Ob}\, \mathcal{C}$, and given

$$\begin{array}{ccc} F & & G \\ \downarrow & & \downarrow \\ \mathrm{Ob}\, \mathcal{C} & \xrightarrow{\ f\ } & \mathrm{Ob}\, \mathcal{C}' \end{array}$$

we have $(f^*G)(X) = G(fX)$, so f^*G is just the usual pull back of G via f. Furthermore $f_! F = F$ is viewed as a set over $\mathrm{Ob}\, \mathcal{C}'$ via f since $f_!(F)(Y) = \coprod_{X \in f^{-1}Y} F(X)$, and this is just Kan's formula

---

[1] For $f_!$ this formula also appears in SGA4 Éxp. 1 by Grothendieck and Verdier [6].

$$(f_! F)(Y) = \varinjlim_{(X, fX \to Y)} F(X) \text{ but a discrete category so OK.}$$

As to f_*, we have

$$\mathrm{Hom}_{\widehat{C'}}(G, f_* F) = \mathrm{Hom}_{\widehat{C}}(f^* G, F)$$

but

$$\mathrm{Hom}(G, f_* F) = (f_* F)(Y) \quad \text{with } f^* G = f^{-1} Y \subset \mathrm{Ob}\, C$$

and

$$\mathrm{Hom}_{\text{sets over Ob}\, C}(f^* G, F) = \text{sections of } F \text{ over } f^{-1} Y$$
$$= \prod_{X \in f^{-1} Y} F(X).$$

In this case

$$(f_! F)(Y) = \coprod_{X \in f^{-1} Y} F(X),$$

$$(f_* F)(Y) = \prod_{X \in f^{-1} Y} F(X).$$

Example 9.5 Suppose $C = \widetilde{G}$ and $C' = \widetilde{G}'$ with $f : G \to G'$ a group homomorphism. Then $\widehat{C}$ is the category of G-sets, $\widehat{C'}$ the category of G'-sets, and

$$(G\text{-sets}) \;\; \substack{\xrightarrow{\;f_!\;} \\ \xleftarrow{\;f^*\;} \\ \xrightarrow{\;f_*\;}} \;\; (G'\text{-sets})$$

where essentially $f^* S = S \circ f$ and $f^* S = S$ viewed as a G-set via f. Furthermore, we have

$$f_! S = G' \times_G S \overset{d}{=} (G \times S)/(g', gs) \sim (g' f(g), s)$$

and in fact

$$\mathrm{Hom}_{G'-\text{set}}(G' \times_G S, T) = \mathrm{Hom}_{G\text{-set}}(S, f^* T).$$

We also have that

$$f_* S = \mathrm{Hom}_{G\text{-set}}(G', S) = \{h : G' \to S \mid h(f(g)g') = gh(g')\}$$

is a G'-set with G' acting by

$$(g'h)(g_1') = h(g_1' g'),$$

in analogy with the case of rings.

The Kan formulae are

$$(f_! F)(Y) = \varinjlim_{(X, fX \xrightarrow{u} Y)} F(X) = \mathrm{Coker}\left\{ \coprod_{Y \xleftarrow{} fX_0} F(X_0) \Leftarrow \coprod_{Y \xleftarrow{u} fX_0 \xleftarrow{fv} fX_1} F(X_1) \right\}$$

where

$$\mathrm{Coker}\left\{ A \overset{d_0}{\underset{d_1}{\Leftarrow}} B \right\} = A/(d_0 b \sim d_1 b).$$

For the case of groups

$$f_! S(*') = \mathrm{Coker}\left\{ \coprod_{*' \xleftarrow{u \in G'} *'} S(*) \Leftarrow \coprod_{* \xleftarrow{u' \in G'} *' \xleftarrow{v \in G} *} S(*) \right\}$$

$$= \mathrm{Coker}\left\{ G' \times S \Leftarrow G' \times G \times S \right\}$$

$$(uf(v), s) \leftarrow\!\!\mid (u, v, s)$$

$$(u, vs) \leftarrow\!\!\mid$$

$$= G' \times_G S.$$

As an exercise, check Kan also for f_*, i.e., confirm that

$$\varprojlim_{X, Y \xrightarrow{u} fX} = \mathrm{Hom}_{G'}(G, S).$$

Chapter 10
Abelian and Additive Categories

Abelian Categories

For example, $\mathrm{Mod}_A = $ (left) A-modules or functors $\mathrm{Funct}(\mathcal{C}, \mathrm{Mod}_A)$ with $\mathcal{C}$ a small category.

In Mod_A, we have that $\mathrm{Hom}_A(M, N)$ is an abelian group in a natural way and have a notion of exact sequences. We have the same structures in functors $\mathrm{Funct}(\mathcal{C}, \mathrm{Mod}_A)$, namely

$$\mathrm{Hom}_{\mathrm{Funct}}(F, G) = \mathrm{Ker}\left\{ \prod_{X \in \mathrm{Ob}\,\mathcal{C}} \mathrm{Hom}_A(F(X), G(X)) \rightrightarrows \prod_{\substack{X \xrightarrow{u} Y \\ \mathrm{in}\,\mathcal{C}}} \mathrm{Hom}_A(F(X), G(Y)) \right\},$$

where the sum of $\theta : F \to G$ and $\eta : F \to G$ is $(\theta + \eta)(X) = \theta(X) + \eta(X)$. Moreover, a sequence of functors $F \to F' \to F''$ from $\mathcal{C}$ to Mod_A is *exact* when $F(X) \to F'(X) \to F''(X)$ is exact for all $X \in \mathrm{Ob}\,\mathcal{C}$.

Question Can we get at the abelian group structure on $\mathrm{Hom}_A(M, N)$ more intrinsically in terms of the category Mod_A?

If X is an object of $\mathcal{C}$, then a *group* (respectively *ring, lattice*, etc.) *structure* on X is a group (etc.) structure on the functor $h_X = \mathrm{Hom}_{\mathcal{C}}(\cdot, X)$, i.e., for all $Y \in \mathrm{Ob}\,\mathcal{C}$ we specify a group (etc.) structure on $\mathrm{Hom}_{\mathcal{C}}(Y, X)$ so that for all $u : Y \to Y'$ in $\mathcal{C}$, we have a group (etc.) homomorphism $\mathrm{Hom}_{\mathcal{C}}(Y', X) \to \mathrm{Hom}_{\mathcal{C}}(Y, X)$. Put still differently, a group structure on X is a lifting

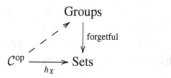

of h_X to a functor $\mathcal{C}^{\mathrm{op}} \to$ Groups. Suppose now $\mathcal{C}$ has a terminal object e and that the products $X \times X$ and $X \times X \times X$ exist. Note that the group law gives $h_X \times h_X \to h_X$, i.e.,

© Springer Nature Switzerland AG 2020
R. Penner, *Topology and K-Theory*, Lecture Notes in Mathematics 2262,
https://doi.org/10.1007/978-3-030-43996-5_10

$$\text{Hom}(Y, X) \times \text{Hom}(Y, X) \xrightarrow{\text{group mult}} \text{Hom}(Y, X)$$

$$\uparrow \simeq$$

$$\text{Hom}(Y, X \times X)$$

where the vertical map is given by $f \mapsto (\text{pr}_1 f, \text{pr}_2 f)$, and so by Yoneda's lemma, this natural transformation $h_{X \times X} \to h_X$ is given by a morphism $m : X \times X \to X$.

The unit element gives a natural transformation $\text{Hom}(Y, e) = \text{point} \to \text{Hom}(Y, X)$, and by Yoneda's lemma, this comes from $e \xrightarrow{1} X$.

Finally and similarly, the group inverse is given by a map $X \xrightarrow{\mu} X$.

Then the group axioms correspond to the following diagrams commuting:

$$
\begin{array}{ccc}
X \times X \times X & \xrightarrow{\;\text{id} \times m\;} & X \times X \\
\downarrow{\scriptstyle m \times \text{id}} & & \downarrow{\scriptstyle m} \\
X \times X & \xrightarrow{\;\;m\;\;} & X
\end{array}
$$

$$
\left.
\begin{array}{l}
X \xrightarrow{(\text{id},1)} X \times X \xrightarrow{m} X \\
a \longmapsto \;\;(a, 1)\;\; \longmapsto a1 \\[2mm]
X \xrightarrow{(1,\text{id})} (X \times X) \xrightarrow{m} X \\
a \longmapsto \;\;(1, a)\;\; \longmapsto 1a
\end{array}
\right\}
$$
are id_X, where $1 : X \to X$ is short for $X \xrightarrow{\text{unique}} e \xrightarrow{1} X$

$$X \xrightarrow{(\text{id},\mu)} X \times X \xrightarrow{m} X$$
$$1$$

$$X \xrightarrow{(\mu,\text{id})} X \times X \longrightarrow X$$
$$1$$

In Mod_A, every object N is an abelian group object in a natural way because $\text{Hom}_A(\cdot, N)$ has a natural abelian group structure.

A *cogroup structure* on X in $\mathcal{C}$ is the same as a group structure on X in $\mathcal{C}^{\text{op}}$, i.e., a group structure on $\text{Hom}(X, \cdot) : \mathcal{C} \to \text{Sets}$. Such a structure is given by

$$X \longrightarrow X \coprod X, \text{ comultiplication,}$$
$$X \longrightarrow \phi, \; \phi = \text{initial object,}$$
$$X \longrightarrow X, \text{ an inverse}$$

satisfying the appropriate diagrams.

Thus, in Mod_A, every object is naturally a co-abelian group object since $\text{Hom}_A(M, \cdot)$ is naturally an abelian group.

Example Any free group is a co-group in the category of groups since

$$\text{Hom}_{\text{Groups}}(F(S), G) = \prod_S G,$$

where $F(\bullet)$ denotes free group on $\bullet$.

Exercise Interpret the arrows above in this setting.

Theorem *Suppose we have a monoid object X and a co-monoid object Y in a category $\mathcal{C}$. Then the two possible monoid structures on $\text{Hom}_{\mathcal{C}}(Y, X)$ coincide and this unique operation is abelian.*

Standard example $\mathcal{C} = $ category of spaces with basepoint and homotopy classes of basepoint-preserving maps. Take G to be an H-space (e.g., $G = \Omega X$, the loop space) and Y to be S^1, which is a co-group because homotopy classes of maps $[S^1, X] = \pi_1 X$. Then the theorem says the two operations on $[S^1, G] = \pi_1 G$ coincide and are abelian.

Proof of Theorem Denote

$$Y \xrightarrow{\gamma} Y \coprod Y, \text{ comultiplication}$$
$$Y \xrightarrow{\varepsilon} \phi, \text{ co-unit}$$
$$\widetilde{\varepsilon} : Y \xrightarrow{\varepsilon} \phi \xrightarrow{\text{unique}} Y$$

Thus we have

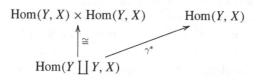

Apply the monoid-valued functor $\text{Hom}(\cdot, X)$ to these arrows

$$\text{Hom}(Y, X) \times \text{Hom}(Y, X) \qquad \text{Hom}(Y, X)$$

$$\uparrow \cong \qquad \nearrow \gamma^*$$

$$\text{Hom}(Y \coprod Y, X)$$

We thus get a map $M \times M \xrightarrow{\theta} M$, where M is the monoid $\text{Hom}(Y, X)$ with operation coming from $X \times X \to X$. θ is a monoid homomorphism so that $\theta(m, 1) = m = \theta(1, m)$ since ε behaves as a co-unit for γ. Now,

$$\theta(m_1, m_2) = \theta((m_1, 1)(1, m_2)) = \theta(m_1, 1)\,\theta(1, m_2) = m_1 m_2 \,,$$

so θ has same effect as the product on M. Remains to check abelian. To see this, replace $X \times X \xrightarrow{m} X$ by $X \times X \xrightarrow{\text{flip}} X \times X \xrightarrow{m} X$. $\qquad\square$

In Mod_A, we have $\phi = e$ since the 0 module is both initial and final. Also, the canonical map $M \coprod N \xrightarrow{\alpha} M \times N$ is an isomorphism, where α is given by

$$\text{pr}_1\,\alpha\,\text{in}_1 = \text{id}_M\,,\quad \text{pr}_1\,\alpha\,\text{in}_2 = 0\,,$$
$$\text{pr}_2\,\alpha\,\text{in}_1 = 0\,,\quad \text{pr}_2\,\alpha\,\text{in}_2 = \text{id}_N\,.$$

Moreover, the co-monoid (in fact, co-abelian group) structure on M is given by

$$M \xrightarrow{\Delta} M \times M \xleftarrow[\cong]{\alpha} M \coprod M \,,$$

and the abelian group structure is given by

$$M \times M \xrightarrow{\alpha^{-1}} M \coprod M \xrightarrow{\text{fold}} M \,.$$

We are led to define an *additive category* as a category $\mathcal{A}$ so that

(1) $\mathcal{A}$ has an object 0 which is both initial and final,
(2) for any $M, N \in \text{Ob}\,\mathcal{A}$, $M \coprod N$ and $M \times N$ exist,
(3) the canonical map $M \coprod N \xrightarrow{\alpha} M \times N$ is an isomorphism,
(4) $\text{Hom}_{\mathcal{A}}(M, N)$ is an abelian group for all $M, N \in \text{Ob}\,\mathcal{A}$, with $\text{Hom}_{\mathcal{A}}(\cdot, \cdot)$ given the natural structure as below.
    Notice that (2) and (3) imply that $\text{Hom}_{\mathcal{A}}(M, N)$ is an abelian monoid with $f + g$ defined as

$$M \xrightarrow{\Delta} M \times M \xrightarrow{\alpha^{-1}} M \coprod M \xrightarrow{(f,g)} N$$

or equivalently as

$$M \xrightarrow{(f,g)} N \times N \xrightarrow{(\text{fold})\circ\alpha^{-1}} N \,.$$

We denote it as $M \oplus N = M \coprod N \approx M \times N$.
For $\mathcal{A}$ an additive category and $f : M \to N$ in $\mathcal{A}$, we define a *cokernel* for f as a pair C, p, where $p : N \to C$ in $\mathcal{A}$ so that for all objects X

$$0 \longrightarrow \text{Hom}(C, X) \xrightarrow{p^*} \text{Hom}(N, X) \xrightarrow{f^*} \text{Hom}(M, X)$$

is exact, i.e., $u \circ f$ zero implies u factors through C

$$M \xrightarrow{f} N \xrightarrow{p} C$$

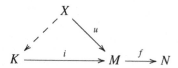

Dually, a *kernel* for $f : M \to N$ is a map $i : K \to M$ so that for all X

$$0 \longrightarrow \mathrm{Hom}(X, K) \xrightarrow{i_*} \mathrm{Hom}(X, M) \xrightarrow{f_*} \mathrm{Hom}(X, N)$$

is exact, i.e., $f \circ u$ zero implies u factors through K

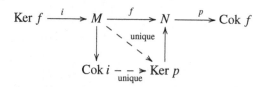

Now given f, take the kernel and cokernel to get

$$\mathrm{Ker}\, f \xrightarrow{i} M \xrightarrow{f} N \xrightarrow{p} \mathrm{Cok}\, f$$

with the diagram showing the unique map $\mathrm{Cok}\, i \dashrightarrow \mathrm{Ker}\, p$.

Ker p is called the *image* Im f and Cok i the *coimage* Coim f of f.

We define an *abelian category* as an additive category $\mathcal{A}$ so that

(5) for any map f, Ker f and Cok f exist,

(6) for any map f, the canonical map Coim $f \to$ Im f is an isomorphism.

Chapter 11
Diagram Chasing in Abelian Categories

Example (*standard example of additive but non-abelian category*) Let $\mathcal{A}$ be the category of topological abelian groups. Then $\mathrm{Hom}_{\mathcal{A}}(A, B)$ is itself an abelian group. Checking Condition (5), suppose $f : A \to B$, then $\mathrm{Ker}\, f$ is the usual set-theoretic kernel with the topology induced by A. $\mathrm{Cok}\, f$ is the group $B/f(A)$ with the quotient topology, so the topology is Hausdorff for $f(A)$ closed. Condition (6) breaks down however: the map

$$\mathbb{R}_{\text{discrete}} \xrightarrow{\text{id}} \mathbb{R}_{\text{usual}}$$

from $\mathbb{R}$ with the discrete topology to $\mathbb{R}$ with the usual topology has $\mathrm{Ker} = \mathrm{Cok} = 0$, while $\mathrm{Im} = \mathbb{R}_{\text{usual}}$, $\mathrm{Coim} = \mathbb{R}_{\text{discrete}}$, i.e., $\mathrm{Im} = f(A)$ has the subspace topology and $\mathrm{Coim}\, f = f(A) \cong A/\mathrm{Ker}\, f$ has the quotient top.

Example If $\mathcal{C}$ is a small category, then $\mathcal{A} = \mathrm{Funct}\,(\mathcal{C}, \mathrm{Ab})$ is an abelian category. We can think of functors as diagrams of abelian groups indexed by $\mathcal{C}$.

Suppose $u : F \to G$ is a natural transformation, then $\mathrm{Ker}\, u$ is computed as

$$(\mathrm{Ker}\, u)(X) = \mathrm{Ker}\,(u(X) : F(X) \longrightarrow G(X)),$$

and similarly,

$$(\mathrm{Cok}\, u)(X) = \mathrm{Cok}\left(F(X) \xrightarrow{u(X)} G(X)\right).$$

As for checking Condition (6), we have

$$\mathrm{Ker}\, u(X) \longrightarrow F(X) \xrightarrow{u(X)} G(X) \longrightarrow \mathrm{Cok}\, u(X).$$

$$\mathrm{Coim}\, u(X) \longrightarrow \mathrm{Im}\, u(X)$$

R. Penner, *Topology and K-Theory*, Lecture Notes in Mathematics 2262,
https://doi.org/10.1007/978-3-030-43996-5_11

Note that sheaves of abelian groups on topological spaces is also an abelian category, but this requires some work that we shall skip.

Let $\mathcal{A}$ be an abelian category. Then $f : X \to Y$ is an *epimorphism* if Cok $f = 0$ (i.e., if for all T, Hom$(Y, T) \hookrightarrow$ Hom(X, T)), and is a *monomorphism* if Ker $f = 0$ (i.e., for all T, Hom$(T, X) \hookrightarrow$ Hom(T, Y)). The adjectival form of epimorphism is *epic* and of monomorphism is *monic*. If f is epic, then we write $A \twoheadrightarrow B$, and if f is monic, then we write $A \hookrightarrow B$.

If f is epic, then we have

$$\text{Ker } f \longrightarrow X \xrightarrow{\ f\ } Y \longrightarrow 0$$

$$\text{Coim } f =\!=\!= \text{Im } f$$

so that $Y = \text{Cok}(\text{Ker } f \longrightarrow X)$.

This has the consequence that if we are given $X \xrightarrow{\ f\ } Y$, in order to construct $Y \to T$, it suffices to construct $X \to T$ that carries Ker f to zero. This is the basis of the

General Principle Any of the usual diagram-chasing arguments used for modules can be extended to an arbitrary abelian category.

Below we indicate a proof of the Serpent Lemma in an arbitrary abelian category to illustrate how this is done.

Proposition *(i) The pull back of an epic $f : Y \to Z$ by any map $X \to Y$ is again epic, i.e.,*

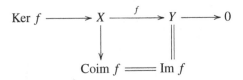

$$f \text{ epic implies } f' \text{ epic.}$$

(ii) Dually, the push out of a monic $f : A \to B$ by any map $A \to C$ is again monic, i.e.,

$$
\begin{array}{ccc}
A & \xrightarrow{\ g\ } & C \\
f \downarrow & & \downarrow f' \\
B & \xrightarrow[\ g'\ ]{} & B +_A C
\end{array}
$$

$$f \text{ monic implies } f' \text{ monic.}$$

Proof C is abelian if and only if C^{op} is abelian, so we prove only (ii).

We define $B +_A C = \mathrm{Cok}\left(A \xrightarrow{(f,-g)} B \oplus C\right)$. Then by definition of Cok, we have an exact sequence

$$0 \longrightarrow \mathrm{Hom}(B +_A C, T) \longrightarrow \mathrm{Hom}(B \oplus C, T) \xrightarrow{(f,-g)^*} \mathrm{Hom}(A, T)$$

$$\mathrm{Hom}(B, T) \oplus \mathrm{Hom}(C, T)$$

$$f^* - g^*$$

Thus, a map $B +_A C \to T$ is the "same" as maps $B \to T$ and $C \to T$ which agree on A, and so $B +_A C$ is the usual push out.

Suppose now that $A \xrightarrow{g} C$ is monic. Then we claim $A \xrightarrow{(f,-g)} B \oplus C$ is monic. Suppose we have $T \longrightarrow A \longrightarrow B \oplus C \longrightarrow C$ so $T \to C$ is the zero map

and $A \to C$ is monic which implies that $T \to A$ is zero, proving the claim.

We proved before that $X \to Y$ epic implies $Y = \mathrm{Cok}\,(\mathrm{Ker}\,f \to X)$, and dually for $A \to B \oplus C$ monic, we have $A = \mathrm{Ker}\,\{B \oplus C \to B +_A C\}$. We thus have the commutative diagram

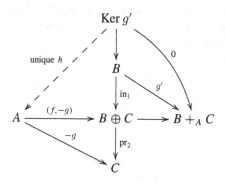

But $-g$ monic and $-g \circ h = 0$ implies $h = 0$, so $\mathrm{Ker}\,g' \to B$ is the zero map whence $\mathrm{Ker}\,g' = 0$, as desired.   □

To substantiate the general principle above, we indicate a proof of the serpent lemma in an abelian category $\mathcal{A}$.

Serpent Lemma *Given a commutative diagram*

$$A' \longrightarrow A \longrightarrow A'' \longrightarrow 0$$

$$\downarrow u' \qquad \downarrow u \qquad \downarrow u''$$

$$0 \longrightarrow B' \longrightarrow B \longrightarrow B''$$

Then there exists an exact sequence

$$\mathrm{Ker}\, u' \longrightarrow \mathrm{Ker}\, u \longrightarrow \mathrm{Ker}\, u'' \overset{\delta}{\longrightarrow} \mathrm{Cok}\, u' \longrightarrow \mathrm{Cok}\, u \longrightarrow \mathrm{Cok}\, u''.$$

The interesting part of the proof is the construction of δ, which we next discuss.

Take the pull back

$$A \times_{A''} \mathrm{Ker}\, u'' \overset{\xi}{\longrightarrow} \mathrm{Ker}\, u''$$

$$\downarrow \qquad\qquad\qquad \uparrow$$

$$A \longrightarrow A''$$

and by the usual diagram chase, we get a map

$$A \times_{A''} \mathrm{Ker}\, u'' \overset{\partial}{\longrightarrow} \mathrm{Cok}\, u'.$$

Now, $\mathrm{Ker}\, u'' = \mathrm{Coker}\,(\mathrm{Ker}\, \xi \to A \times_{A''} \mathrm{Ker}\, u'')$ so by the remark before the general principle above, to get $\delta : \mathrm{Ker}\, u'' \to \mathrm{Cok}\, u'$ it suffices to check that $\mathrm{Ker}\, \xi$ maps to zero under ∂. To see this, let $C = \mathrm{Ker}\,(A \to A'')$ and note that a diagram chase gives $A \times_{A''} \mathrm{Ker}\, u'' \to C$. Take the pull back

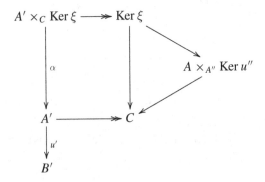

Thus given $x \in \mathrm{Ker}\, \xi$, pull back to $y \in A' \times_c \mathrm{Ker}\, \xi$. Then $u' \circ \alpha(y)$ is the coset of $\partial(x)$, i.e., $\partial(\mathrm{Ker}\, \xi) = 0$, as desired.

Exercise Finish the proof of the serpent lemma.   □

Chapter 12
Fibered and Cofibered Categories

Suppose that $f : \mathcal{C} \to \mathcal{C}'$ is a functor and Y is an object in $\mathcal{C}'$. Then we have already discussed the following categories:

(1) the fiber $f^{-1}Y$ of f over Y is the subcategory of $\mathcal{C}$ where arrows lie over id_Y;
(2) the left fiber f/Y over Y whose objects are (X, u), where $X \in \mathrm{Ob}\,\mathcal{C}$ and $u : f(X) \to Y$, and a map $(X, u) \to (X', u')$ is given by maps $X \xrightarrow{v} X'$ so that the diagram

$$\begin{array}{ccc} f(X) & \xrightarrow{\ u\ } & Y \\ {\scriptstyle f(v)}\downarrow & \nearrow {\scriptstyle u'} & \\ F(X') & & \end{array}$$

  commutes;
(3) the right fiber $Y \backslash f$ consists of $(X, u : Y \to f(X))$.

With $\widehat{\mathcal{C}} = \mathrm{Funct}(\mathcal{C}, \mathrm{Sets})$, for $\mathcal{C}$ small, f induces $f^* : \widehat{\mathcal{C}'} \to \widehat{\mathcal{C}}, G \mapsto G \circ f$.

Proposition *We have adjoint functors*

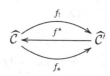

that is,

(i) $\mathrm{Hom}(f_! F, G) = \mathrm{Hom}(F, f^*G)$,
(ii) $\mathrm{Hom}(G, f_* F) = \mathrm{Hom}(f^*G, F)$,

given by Kan's formulae

© Springer Nature Switzerland AG 2020
R. Penner, *Topology and K-Theory*, Lecture Notes in Mathematics 2262,
https://doi.org/10.1007/978-3-030-43996-5_12

$$f_!(F)(Y) \qquad \varinjlim_{X,fX \to Y \in f/Y} F(X),$$

$$f_*(F)(Y) = \varprojlim_{X,Y \to fX \in Y \setminus f} F(X).$$

Proof (i) Start with $\theta : F \to f^*G$,

$$\theta = (\theta_X : F(X) \to G(fX)).$$

Let $\mathcal{M}(f) = \mathcal{C} \times_{\mathcal{C}'} \text{Arr } \mathcal{C}'$, where $\text{Arr } \mathcal{C}'$ are the arrows (that is, the morphisms) in $\mathcal{C}'$, denote the category whose objects are given by pairs $(X, u : fX \to Y)$ and whose maps $(X, u) \to (X', u')$ are a pair $X \xrightarrow{a} X'$ and $Y \xrightarrow{b} Y'$ so that

$$
\begin{array}{ccc}
f(X) & \xrightarrow{u} & Y \\
{\scriptstyle f(a)} \downarrow & & \downarrow {\scriptstyle b} \\
f(X') & \xrightarrow{u'} & Y'
\end{array}
\qquad \text{commutes.}
$$

Given $(X, u : f(X) \to Y)$ in $\mathcal{M}(f)$, we define

$$\theta_{(X,u)} : F(X) \xrightarrow{\theta_X} G(fX) \xrightarrow{G(u)} G(Y).$$

Thus $\theta = (\theta_X)$ gives a family of maps $F(X) \to G(Y)$, for each (X, u), which are natural with respect to maps in $\mathcal{M}(f)$.

Now fix Y. Then the maps $\theta_{(X,u)}$ induce a map

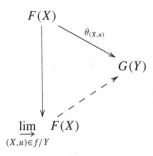

i.e., induce a map $f_! F(Y) \to G(Y)$. We check easily that this gives a natural transformation $f_! F \to G$.

It is clear that the following gadgets are the same:

(1) a natural transformation $\theta : F \to f^*G$ over $\mathcal{C}$,
(2) a natural transformation $F(X) \to G(Y)$ over $\mathcal{M}(f)$,
(3) a natural transformation $f_! F(Y) \to G(Y)$ over $\mathcal{C}'$.

Therefore,

$$\mathrm{Hom}(F, f^*G) = \mathrm{Hom}(f_! F, G).$$

The proof of (ii) is exactly dual.   □

Fibered Category [due to Grothendieck]

Example Let $\mathcal{B}$ be the category of topological spaces. For each $X \in \mathrm{Ob}\,\mathcal{B}$, we let $\mathcal{E}_X$ be the *category of spaces over* X i.e., pairs E, $p : E \to X$ in $\mathcal{B}$. This is denoted $\mathcal{B}/X$. Now, if $u : X \to X'$ is a map in $\mathcal{B}$, we have a pull back functor $u^* : \mathcal{E}_{X'} \to \mathcal{E}_X$

$$\begin{pmatrix} E \\ \downarrow \\ X' \end{pmatrix} \longmapsto \begin{pmatrix} X \times_{X'} E \\ \downarrow \\ X \end{pmatrix}.$$

Given $Z \xrightarrow{v} Y \xrightarrow{u} X$ in $\mathcal{B}$, we have a canonical isomorphism

$$Z \times_Y (Y \times_X E) \simeq Z \times_X E$$

given by $c_{u,v} : v^* u^* \xrightarrow{\;\sim\;} (uv)^*$.

We have the following *(cocycle) condition*

$$
\begin{array}{ccc}
w^* v^* u^* & \xrightarrow{\;w^*(c_{u,v})\;} & w^*(uv)^* \\
\downarrow{\scriptstyle (c_{v,w})u^*} & & \downarrow{\scriptstyle c_{uv,w}} \\
(v\,w)^* u^* & \xrightarrow[\;c_{u,vw}\;]{} & (uvw)^*
\end{array}
$$

where the diagram commutes.

First definition of *fibered category* over a category $\mathcal{B}$: It is a *pseudo-functor* from $\mathcal{B}^{\mathrm{op}}$ to the category Cat of categories. Namely, a way of assigning a category $\mathcal{E}_X$ to each $X \in \mathrm{Ob}\,\mathcal{B}$, to each arrow $u : Y \to X$ a functors $u^* : \mathcal{E}_X \to \mathcal{E}_Y$ and to each pair $Z \xrightarrow{v} Y \xrightarrow{u} X$ in $\mathcal{B}$ an isomorphism of functors $c_{u,v} : v^* u^* \xrightarrow{\;\sim\;} (uv)^*$ so that the cocycle condition holds. Note that a functor from $\mathcal{B}$ to Cat is a pseudo-functor so that $c_{u,v}$ is the identity.

New definition of *fibered category*: Suppose we have a functor $p : \mathcal{E} \to \mathcal{B}$. If $X \in \mathrm{Ob}\,\mathcal{B}$, then let $\mathcal{E}_X$ denote the fiber $p^{-1}X$. Let $E' \xrightarrow{f} E$ be a morphism in $\mathcal{E}$ lying over $X' \xrightarrow{u} X$, i.e., $p(E') = X'$, $p(E) = X$ and $p(f) = u$. We say that f is *cartesian* provided that for any $F \in \mathrm{Ob}\,\mathcal{E}_{X'}$, we have the isomorphism

$$
\mathrm{Hom}_{\mathcal{E}_{X'}}(F, E') \xrightarrow{\;\sim\;} \{g \in \mathrm{Hom}_{\mathcal{E}}(F, E) : p(g) = u\}
$$
$$
\alpha \longmapsto f\alpha.
$$

Given $p : \mathcal{E} \to \mathcal{B}$, we say $\mathcal{E}$ is a *pre-fibered category* over $\mathcal{B}$ if given any $u : X' \to X$ in $\mathcal{B}$ and object $E \in \mathcal{E}_X$, there exists a *cartesian arrow* $E' \to E$ lying over u. A *fibered category* is a pre-fibered category in which the composition of two cartesian arrows is again cartesian.

Suppose $p : \mathcal{E} \to \mathcal{B}$ is pre-fibered. Then given any $u : X' \to X$, we get a functor $u^* : \mathcal{E}_X \to \mathcal{E}_{X'}$ by choosing for each $E \in \mathcal{E}_X$ a cartesian arrow $E' \to E$ over u and setting $u^*E = E'$. Then

$$\mathrm{Hom}_{\mathcal{E}_{X'}}(f, u^*E) \simeq \{g : F \to E : p(g) = u\},$$

$$\mathrm{Hom}_{\mathcal{E}}(F, E)_u \overset{d}{=} \{g : F \to E : p(g) = u\},$$

and then for all F over X',

$$\mathrm{Hom}_{\mathcal{E}_{X'}}(F, u^*E) \overset{\sim}{\longrightarrow} \mathrm{Hom}_{\mathcal{E}}(F, E)_u .$$

Therefore u^*E is unique up to canonical isomorphism.

If $X'' \overset{v}{\to} X' \overset{u}{\to} X$, then there is a canonical arrow

$$v^*u^* \longrightarrow (uv)^*,$$

and we can check it satisfies the co-cycle condition and is an isomorphism $\mathcal{E} \to \mathcal{B}$ in a fibered category. This indicates that the two definitions are the same.

Dual Notions $p : \mathcal{E} \to \mathcal{B}$ is *pre-cofibered* when for every $u : X \to Y$ in $\mathcal{B}$ we have a functor $u_* : \mathcal{E}_X \to \mathcal{E}_Y$ so that

$$\mathrm{Hom}_{\mathcal{E}_Y}(u_*E, F) \simeq \mathrm{Hom}_{\mathcal{E}}(E, F)_u .$$

Cofibered means pre-cofibered and composition of co-cartesian is co-cartesian.

Now go back to $f : \mathcal{C} \to \mathcal{C}'$ with Y and object in $\mathcal{C}'$ where

$$f_!(F)(Y) = \varinjlim_{(X, u : fX \to Y) \in f \backslash Y} F(X) .$$

Claim 12.1 *If f is pre-cofibered, then*

$$f_!(F)(Y) = \varinjlim_{X \in f^{-1}Y} F(X) .$$

Moreover,

$$f_*(F)(Y) = \varprojlim_{\left(X, Y \overset{u}{\to} fX\right) \in Y \backslash f} F(X) .$$

Claim 12.2 *If f is pre-fibered, then*

$$f_*(F)(Y) = \varprojlim_{X \in f^{-1}Y} F(X).$$

Exercise Prove the two claims.

Chapter 13
Examples of Fibered Categories

In a pre-cofibered category $\mathcal{C} \xrightarrow{f} C$, we have maps

$$i \; : \; f^{-1}Y \hookrightarrow f/Y$$
$$: E \longmapsto \begin{pmatrix} E \\ fE \overset{\text{id}}{=} Y \end{pmatrix}$$

and

$$r \; : \; f/Y \longrightarrow f^{-1}Y$$
$$: \begin{pmatrix} X \\ fX \overset{u}{\to} Y \end{pmatrix} \longmapsto u_*X \, ,$$

and it is clear that $ri = \mathrm{id}_{f^{-1}Y}$.

Claim *That r is left adjoint to i.*

Proof

$$\mathrm{Hom}_{f/Y} \left(\begin{pmatrix} X \\ fX \overset{u}{\to} Y \end{pmatrix}, \begin{pmatrix} E \\ fE = Y \end{pmatrix} \right)$$
$$= \mathrm{Hom}_{\mathcal{C}}(X, E)_u \quad \text{by definition}$$
$$= \mathrm{Hom}_{f^{-1}Y}(u_*X, E) \text{ by the universal property of } u_*X. \qquad \square$$

Exercise Check that f is pre-cofibered if and only if $f^{-1}Y \hookrightarrow f/Y$ has a left adjoint for all $Y \in \mathcal{C}'$.

© Springer Nature Switzerland AG 2020
R. Penner, *Topology and K-Theory*, Lecture Notes in Mathematics 2262,
https://doi.org/10.1007/978-3-030-43996-5_13

Claim

$$\varinjlim_{\left(X, fX \xrightarrow{u} Y\right)} F(X) = \varinjlim_{X \in f^{-1}Y} FX.$$

Proof Consider $A \underset{i}{\overset{r}{\rightleftarrows}} B$ with $ri = \mathrm{id}_A$ and r a left adjoint to i, so there is a natural transformation $\mathrm{id}_B \to ir$, and suppose $B \xrightarrow{F}$ Sets. We show

$$\varinjlim_{B \in B} F(B) = \varinjlim_{A \in A} F(iA).$$

We have a canonical map $\varinjlim_{A \in A} F(iA) \to \varinjlim_{B \in B} F(B)$, and using the universal property of $\varinjlim$, we have to show that

$$\varprojlim_{B \in B} \mathrm{Hom}_{\mathrm{Sets}}(F(B), S) \xrightarrow{\sim} \varprojlim_{A \in A} \mathrm{Hom}_{\mathrm{Sets}}(F(iA), S).$$

Then $G(B) = \mathrm{Hom}_{\mathrm{Sets}}(F(B), S)$ is a contravariant functor, and we want to prove

(∗) $$\varprojlim_{B \in B} G(B) \xrightarrow{\sim} \varinjlim_{A \in A} G(iA).$$

An element of the left-hand side is a family $g_B \in G(B)$ compatible with the arrows in B. That the map in (∗) is an isomorphism follows from the picture

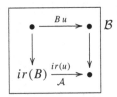

Example 13.1 For $A \overset{i}{\underset{\substack{\| \\ \text{point}}}{\hookrightarrow}} B$, i has a left adjoint if and only if $i(\text{point}) = e \in B$ is a final object for B. Then the claim says $\varinjlim F(B) = F(e)$ for $F : B \to$ Sets.

Example 13.2 This is false for contravariant functors, e.g., take

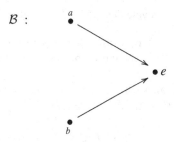

$$\mathcal{B} :$$

Then $\varinjlim F = F(a) \coprod_{F(e)} F(b) \neq F(e)$.

Example 13.3 A *discrete* category has all morphisms identity morphisms, so the arrows comprise a set. Suppose $\mathcal{E} \xrightarrow{p} \mathcal{B}$ is a fibered category so that the fibers are discrete for each X in $\mathcal{B}$. We get, then, a contravariant functor $\mathcal{B}^{op} \to$ Sets, $B \mapsto \mathrm{Ob}\,\mathcal{E}_X$, that is, a pseudo-functor is in fact a functor in a discrete category.

Conversely, given any functor $F : \mathcal{B}^{op} \to$ Sets, we can form the category $\mathcal{B}/F$ whose objects are (X, ξ), $X \in \mathrm{Ob}(\mathcal{B})$, $\xi \in F(X)$, and

$$p : \mathcal{B}/F \longrightarrow \mathcal{B}$$
$$(X, \xi) \longmapsto X$$

is a fibered category with discrete fibers. The fiber over X is the set $F(X)$ regarded as a discrete category.

Note that

$$\mathrm{Hom}_{\mathcal{B}/F}((X, \xi), (Y, \eta)) = \{f : X \to Y : F(f)\eta = \xi\},$$

and in $\mathcal{B}/F$ every arrow is cartesian.

Thus fibered categories over $\mathcal{C}$ with discrete fibers are the same as $F : \mathcal{C}^{op} \to$ Sets, and cofibered categories over $\mathcal{C}$ with discrete fibers are the same as $F : \mathcal{C} \to$ Sets.

Example 13.4 Suppose $U \xrightarrow{p} G$ is a group homomorphism and consider $p : \widetilde{U} \to \widetilde{G}$. If this is fibered, then p must be epic, and every arrow is cartesian by unique cancellation. Thus $\widetilde{U} \to \widetilde{G}$ is a fibered category if and only if $U \to G$ is epic, and the fiber over the unique object of $\widetilde{G}$ is the category $\widetilde{K}$ where K is $\mathrm{Ker}\{p : U \to G\}$.

In general, given a fibered category $p : \mathcal{E} \to \mathcal{B}$, we can define for each $u : X \to Y$ in $\mathcal{B}$ a functor $u^* : \mathcal{E}_Y \to \mathcal{E}_X$, unique up to canonical isomorphism.

In the case of $\widetilde{U} \to \widetilde{G}$, such a choice is the same as a set-theoretic section of U over G. Check that the canonical isomorphisms

$$(uv)^* \xleftarrow{\sim} v^*u^*$$

are the same thing as the 2-cocycle describing U as a group extension.

Example 13.5 Let $f : X \to Y$ be a simplicial map of simplicial complexes. Take C to be the simplexes in X and C' to be the simplexes in Y as posets and let $f : C \to C'$ be the induced map of posets.

If σ is a simplex in Y, then what is f/σ? f/σ is the poset of simplices in the subcomplex $f^{-1}\bar{\sigma}$. Is $f : C \to C'$ fibered? We have

$$
\begin{array}{ccc}
u^* T & \subset & T \\
 & & \updownarrow \\
\sigma' & \underset{u}{\subset} & \sigma
\end{array}
$$

So $u^* T$ is the face of T whose vertices lie over vertices σ'. So yes it is fibered but not cofibered.

Furthermore, $f^{-1}\sigma$ is the poset of all simplexes mapping onto σ, and this is interesting because f^{-1} (open simplex σ) is homeomorphic to $f^{-1}(\xi) \times \sigma$, where $\xi \in \text{Int } \sigma$. So $f^{-1}\xi$ is **not** a simplicial complex, but it is a union of contractible pieces whose inclusion relations are described by the poset $f^{-1}\sigma$.

Chapter 14
Projective Resolutions

Suppose $\mathcal{A} \xrightarrow{F} \mathcal{B}$ is a functor between abelian categories. F is an *additive functor* if $\mathrm{Hom}_{\mathcal{A}}(A, A') \to \mathrm{Hom}_{\mathcal{B}}(FA, FA')$ is a homomorphism of abelian groups.

Example 14.1 $\mathcal{A} = \mathrm{Funct}\,(\mathcal{C}, \mathrm{Ab})$, $\mathcal{B} = \mathrm{Ab}$, with $F = \varinjlim_{\mathcal{C}}$, that is,

$$F(A) = \varinjlim_{X \in \mathcal{C}} A(X)\,.$$

Example 14.2 $\mathcal{A} = G\text{-mod}$, $\mathcal{B} = \mathrm{Ab}$, and

$$F(M) = \mathbb{Z} \otimes_{\mathbb{Z}[G]} M,$$

i.e., the largest quotient on which G acts trivially.

An object P of an abelian category $\mathcal{A}$ is *projective* (respectively *injective*) if $\mathrm{Hom}_{\mathcal{A}}(P, \cdot) : \mathcal{A} \to \mathrm{Ab}$ is exact (respectively $\mathrm{Hom}_{\mathcal{A}}(\cdot, P) : \mathcal{A}^{\mathrm{op}} \to \mathrm{Ab}$ is exact).

In general, given $0 \to A' \to A \to A'' \to 0$ exact in $\mathcal{A}$, we know that for any $P \in \mathcal{A}$ we have

$$0 \longrightarrow \mathrm{Hom}(P, A') \longrightarrow \mathrm{Hom}(P, A) \longrightarrow \mathrm{Hom}(P, A'') \quad \text{exact.}$$

So P is projective if and only if $\mathrm{Hom}(P, \cdot)$ carries epic maps to epic maps.

Similarly, P is injective if and only if $\mathrm{Hom}_{\mathcal{A}}(\cdot, P)$ carries monic maps to epic maps.

If P is projective and we have an epic $A \to P$, then $\mathrm{Hom}(P, A) \twoheadrightarrow \mathrm{Hom}(P, P)$, so there exists $\alpha \mapsto \mathrm{id}_P$ so that P is a direct summand of A and conversely, and therefore P is projective if and only if every $A \twoheadrightarrow P$ has a section.

Thus an R-module P is projective in the category of R-modules if and only if it is a direct summand in a free R-module.

© Springer Nature Switzerland AG 2020
R. Penner, *Topology and K-Theory*, Lecture Notes in Mathematics 2262,
https://doi.org/10.1007/978-3-030-43996-5_14

Suppose $\mathcal{A}$ is an abelian category. A *complex* in $\mathcal{A}$ is a diagram

(∗) $\cdots \xrightarrow{d} A_{n+1} \xrightarrow{d} A_n \xrightarrow{d} A_{n-1} \xrightarrow{d} \cdots ,$

for $n \in \mathbb{Z}$, so that $d^2 = 0$.

The complexes in $\mathcal{A}$ form a category in the obvious way called $C(\mathcal{A})$, and $C(\mathcal{A})$ is an abelian category. $A.$ is the notation for a complex,

$$\mathrm{Ker}\{A. \longrightarrow B.\}_n = \mathrm{Ker}\{A_n \longrightarrow B_n\}$$

and so on.

Now (∗) can also be written

$$\cdots \xrightarrow{d} A^n \xrightarrow{d} A^{n+1} \xrightarrow{d} \cdots$$

where $A^n = A_{-n}$, and we define

$$\begin{aligned}
C_{\geq 0}(\mathcal{A}) &= \text{category of } chain\ complexes \\
&= \{A. \text{ so that } A_n = 0, n < 0\}, \\
C_{\leq 0}(\mathcal{A}) = C^{\geq 0}(\mathcal{A}) &= \text{category of } co\text{-}chain\ complexes, \\
C^{+}(\mathcal{A}) &= \text{category of complexes with } A^n = 0, n \ll 0, \\
C^{-}(\mathcal{A}) &= \text{category of complexes with } A^n = 0, n \gg 0.
\end{aligned}$$

Homology and cohomology are defined as usual.

Suppose we have $f, g : A. \to B.$. Then a *homotopy* between f and g is

$$h = \{h_n : A_n \to B_{n+1}\}$$

so that

$$f - g = dh + hd .$$

Proposition *Homotopic maps induce the same maps on homology.*

The proof is standard and left as an exercise.

Homotopy is an equivalence relation on $\mathrm{Hom}_{C(\mathcal{A})}(A., B.)$, so we can form the abelian group of homotopy classes $[A., B.]$.

Denote $K(\mathcal{A}) = $ category with the same objects as $C(\mathcal{A})$, but in which

$$\mathrm{Hom}_{K(\mathcal{A})}(A., B.) = [A., B.],$$

and define K^{+}, K^{-} as before. Note that the notion of isomorphism in $K(\mathcal{A})$, namely homotopy equivalence, is too strong.

A *quasi-isomorphism* of complexes, or sometimes simply *quis* for short, is a map $A. \to B.$ in $C(\mathcal{A})$ which induces a homology isomorphism.

Example A homotopy equivalence $f : A \to B$ is a quis.

Given $M \in \mathrm{Ob}\mathcal{A}$, a *(left) resolution* of M is an exact sequence of the form

$$\cdots \longrightarrow A_1 \longrightarrow A_0 \longrightarrow M \longrightarrow 0.$$

A *projective resolution* is a resolution so that each A_i is projective.

Key Proposition *Any two projective resolutions of M are homotopy equivalent, and this assignment of projective resolution is functorial in M up to homotopy.*

Again the proof is an exercise.

Think of M as being a complex concentrated in degree 0 and a resolution of M as being a chain complex $A.$ equipped with a quasi-isomorphism

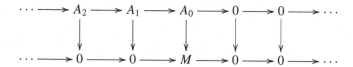

Let $\mathcal{P}$ be the full subcategory of $\mathcal{A}$ consisting of projectives, where a subcategory $\mathcal{B}$ of $\mathcal{A}$ is *full* if for any two of its objects $A, B \in \mathcal{B}$, we have $\mathrm{Hom}_{\mathcal{B}}(A, B) = \mathrm{Hom}_{\mathcal{A}}(A, B)$. $\mathcal{P}$ is an additive category, so it makes sense to talk about $C_+(\mathcal{P})$, $K(\mathcal{P})$, etc.

Define $K(\mathcal{P})$ to be the full subcategory of $K(\mathcal{A})$ consisting of complexes made up of projectives. A *(left) resolution* of a complex $M.$ is another complex $A.$ together with a quis $A. \to M.$, and a *projective resolution* is a resolution $A. \in C(\mathcal{P})$.

Proposition *Provided we stay in $K_+(\mathcal{A})$, i.e., bounded above, any two projective resolutions of M. are isomorphic in $K_+(\mathcal{P})$. Moreover, by choosing for each M a projective resolution $P. \to M.$, we get a well-defined functor $K_+(\mathcal{A}) \to K_+(\mathcal{P})$, and this is adjoint to inclusion, where $M. \mapsto P..$*

Proof *(of the simplest case)* Given

$$
\begin{array}{ccccccccc}
\cdots & \longrightarrow & P_2 & \longrightarrow & P_1 & \longrightarrow & P_0 & \longrightarrow & M & \longrightarrow & 0 \\
& & \downarrow{\scriptstyle u_2} & & \downarrow{\scriptstyle u_1} & & \downarrow{\scriptstyle u_0} & & \downarrow{\scriptstyle u} & & \\
\cdots & \longrightarrow & A_2 & \longrightarrow & A_1 & \longrightarrow & A_0 & \longrightarrow & N & \longrightarrow & 0
\end{array}
$$

where successive horizontal arrows compose to zero, the P_i's are projective and the bottom sequence is exact, there exist $u_0, u_1, u_2, \ldots$ making the diagram commute, and the map of complexes is unique up to homotopy.

A diagram chase gives $u. : M. \to P.$ as usual. Suppose we extend u also to $u'.$, and consider $u. - u'. = \tilde{u}..$ Then we have

$$P_2 \longrightarrow P_1 \longrightarrow P_0 \longrightarrow M \longrightarrow 0$$
$$\downarrow \widetilde{u}_2 \quad \downarrow \widetilde{u}_1 \quad {}_{h_0} \quad \downarrow \widetilde{u}_0 \quad \downarrow 0$$
$$A_2 \longrightarrow A_1 \longrightarrow A_0 \longrightarrow N \longrightarrow 0.$$

Now $\widetilde{u}_0 = d\,h_0$ and $d\,(\widetilde{u}_1 - h_0\,d) = 0$, so we find h_1 with $\widetilde{u}_1 - h_0\,d = d\,h_1$, and so on. This proves the simplest case.

Why are two projective resolutions $P. \to M$, $Q. \to M$ homotopy equivalent? Use the simplest case to get

$$P. \longrightarrow M$$
$$v. \uparrow \downarrow u. \qquad \mathrm{id} \uparrow \downarrow \mathrm{id}$$
$$Q. \longrightarrow M$$

Then

$$P. \longrightarrow M$$
$$v.u. \downarrow \qquad \qquad \downarrow \mathrm{id}$$
$$P. \longrightarrow M$$

so $v.u. \simeq \mathrm{id}_M$ and similarly for $u.v.$. $\qquad\qquad\square$

At this point, we have a functor well defined up to canonical isomorphism given by

$$\mathcal{A} \longrightarrow K_{\geq 0}(\mathcal{P})$$
$$M \longmapsto \text{ projective resolution of } M.$$

Important We must assume that $\mathcal{A}$ *has enough projectives*, i.e., any $M \in \mathrm{Ob}\,\mathcal{A}$ is the quotient of some projective P. Then we construct projective resolutions

$$\ldots \longrightarrow P_1 \longrightarrow P_0 \longrightarrow M \longrightarrow 0.$$

Example The category $\mathcal{A}$ of finite abelian p-groups **does not** have enough projectives (or injectives) since

$$A = \mathbb{Z}/p^r\,\mathbb{Z} \times \cdots \text{ for } A \in \mathcal{A}$$

is **not projective**. Note that $\mathbb{Z}/p^{r+1}\mathbb{Z} \twoheadrightarrow \mathbb{Z}/p^r\,\mathbb{Z}$ is epic yet there is no section.

Chapter 15
Analogues of Homotopy Liftings

Recall last time we constructed $\cdots \to P_1 \to P_0 \to M \to 0$, which is the same as a quis

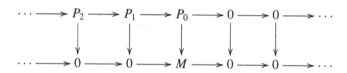

Let

$$\mathcal{A} = \text{abelian category with enough projectives,}$$
$$\mathcal{P} = \text{full subcategory of projectives.}$$

Suppose given a complex $A. \in C_+ \mathcal{A}$; we want to construct a complex and an epimorphism
$$P. \longrightarrow A.$$
which is a quasi-isomorphism with $P. \in C_+(\mathcal{P})$.

Without loss of generality, $A.$ is so that $A_n = 0$ for all $n < 0$. Choose $P_0 \twoheadrightarrow A_0$ and consider the diagram

© Springer Nature Switzerland AG 2020
R. Penner, *Topology and K-Theory*, Lecture Notes in Mathematics 2262,
https://doi.org/10.1007/978-3-030-43996-5_15

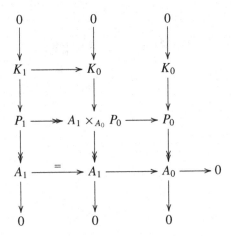

Apply the Serpent Lemma to the first two columns to get $K_1 \twoheadrightarrow K_0$. So far, this gives

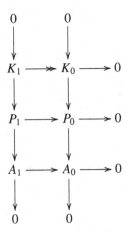

Put

$$K_{10} = \mathrm{Ker}\{K_1 \to K_0\},$$
$$P_{10} = \mathrm{Ker}\{P_1 \to P_0\},$$
$$A_{10} = \mathrm{Ker}\{A_1 \to A_0\}.$$

The Serpent Lemma gives the commutative diagram

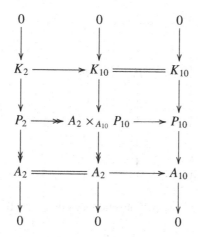

with exact verticals. Apply the Serpent Lemma again to get $K_2 \twoheadrightarrow K_{10}$.

Iterate to get a diagram with exact verticals

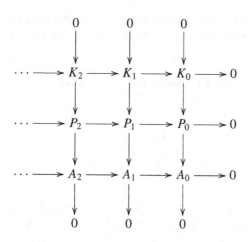

so that $K.$ has trivial homology, whence $H_*P. \simeq H_*A.$, so $P. \to A.$ is indeed a quis.

Another construction: given $A. \in C_+(\mathcal{A})$, we can find a $P. \twoheadrightarrow A.$ so that $P. \in C_+(\mathcal{P})$ and $P.$ is acyclic

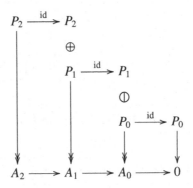

Incidentally, any $P. \in C_{+}\mathcal{P}$ with $P.$ acyclic looks this way. To see this, consider

so $P_1 \approx P_{10} \oplus P_0$ and so on.

The next step using either construction is to show that two projective resolutions of $A. \in C_{+}(\mathcal{A})$ are homotopy equivalent and to prove functoriality in $\mathcal{A}$.

Recall,

Homotopy Extension (HE) Theorem: Suppose given CW complexes A and B, maps

$$
\begin{array}{ccc}
A & & E \\
\cap & & \nearrow \\
i \downarrow & \nearrow f & \\
B & &
\end{array}
$$

and a homotopy $h : A \times I \to E$ so that $h_0 = f i$. Then there exists an extension $\widehat{h} : B \times I \to E$ so that $\widehat{h} = h$ on $A \times I$ and $\widehat{h}_0 = f$, i.e., a lift $-- \to$ making the diagram commute

$$
\begin{array}{ccc}
B \times \{0\} \cup A \times I & \xrightarrow{f \cup h} & E \\
\cap & \nearrow \widehat{h} \nearrow & \downarrow \\
B \times I & \longrightarrow & *
\end{array}
$$

or in other words

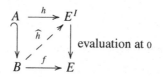

Chain Homotopy (CH) Theorem: Suppose given a fiber space E with

so that

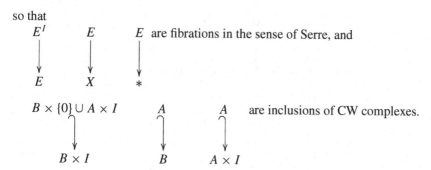

E are fibrations in the sense of Serre, and

are inclusions of CW complexes.

Then each of the following is a homotopy equivalence

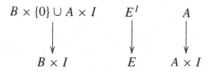

Chain Homotopy Extension (CHE) Theorem: Given

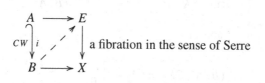

a fibration in the sense of Serre

If either i or p is a homotopy equivalence, then a lift $-\!-\!\to$ exists making the diagram commute.

By analogy, we have

Proposition Suppose given in $C_+(\mathcal{A})$ a diagram

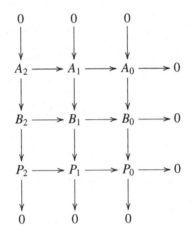

with p epic in each dimension, i monic in each dimension, $B./A. \in C_+(\mathcal{P})$ and either i or p a quis. Then there exists a lift $-- \rightarrow$ making the diagram commute.

Proof Suppose first that p is quis

Let $SK_n(B, A)$ be the complex

$$\cdots \longrightarrow A_{n+1} \longrightarrow A_n \longrightarrow B_{n-1} \longrightarrow B_{n-2} \longrightarrow \cdots$$

Then

$$A. = SK_0(B, A) \subset SK_1(B, A) \subset \cdots \subset SK_n(B, A) \subset \cdots \subset B.$$

are all subcomplexes. Moreover

$$SK_{n+1}(B, A)/SK_n(B, A) = \begin{cases} P_n, & \text{in degree } n, \\ 0, & \text{else.} \end{cases}$$

Thus we have

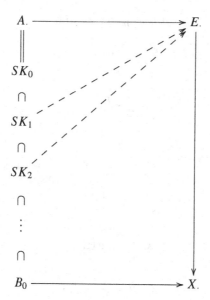

and have thus reduced the problem to the case where $A.$ and $B.$ differ in one degree by a projective object.

Denote $P[n] =$ complex with P in dimension n, and zeroes elsewhere. Because P is projective, $0 \longrightarrow A_n \longrightarrow B_n \longrightarrow P_n \longrightarrow 0$ splits, and so

$$B. = A. \oplus P[n]$$

with $d_n(a, p) = da + \theta p$ where $\theta : P_n \to \mathrm{Ker}\{A_{n-1} \to A_{n-2}\}$. Thus we have

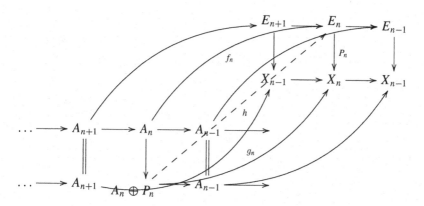

Problem: Define $h : P_n \to E_n$ so that

$$p_n h = g_n \text{ on } P \text{ with}$$

$$dh = f_{n-1}d : A_n \oplus P \to E_{n-1}$$

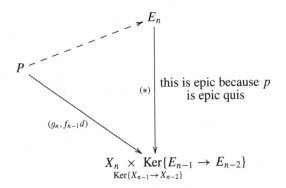

and h exists because $(*)$ is epic and P is projective. □

Chapter 16
The Mapping Cylinder and Mapping Cone

Recall the lifting result from last time which we continue discussing: Given

$$A \xrightarrow{\ f\ } E$$

with i monic, Coker $i \in C_+(\mathcal{P})$, p epic and either i or p a quis, then there exists a lifting h making diagram commute.

We did the case where p is a quis last time.

Now suppose i is a quis. Then Coker $i = (B/A). = P.$, and the long exact sequence gives $H_* P. = 0$. Since $P. \in C_+(\mathcal{P})$, we showed $P.$ is a direct sum of complexes of the form

$$\ldots 0 \longrightarrow 0 \longrightarrow \underset{\deg n}{P} \xrightarrow[\text{id}]{\ \cdot\ } \underset{\deg n-1}{P} \longrightarrow 0 \longrightarrow 0 \longrightarrow \ldots .$$

We have the commutative diagram

© Springer Nature Switzerland AG 2020
R. Penner, *Topology and K-Theory*, Lecture Notes in Mathematics 2262,
https://doi.org/10.1007/978-3-030-43996-5_16

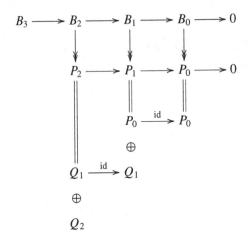

$$
\begin{array}{ccccccccc}
B_3 & \longrightarrow & B_2 & \longrightarrow & B_1 & \longrightarrow & B_0 & \longrightarrow & 0 \\
& & \downarrow & & \downarrow & & \downarrow & & \\
& & P_2 & \longrightarrow & P_1 & \longrightarrow & P_0 & \longrightarrow & 0 \\
& & \| & & \| & & \| & & \\
& & & & P_0 & \xrightarrow{\text{id}} & P_0 & & \\
& & & & & \oplus & & & \\
& & Q_1 & \xrightarrow{\text{id}} & Q_1 & & & & \\
& & \oplus & & & & & & \\
& & Q_2 & & & & & &
\end{array}
$$

Consider the following special case

$$
\begin{array}{ccc}
\{0\} & \xrightarrow{\quad f \quad} & E \\
\downarrow{\scriptstyle i} & \nearrow & \downarrow{\scriptstyle p} \\
\left\{\dots \to 0 \to P_n \xrightarrow{\text{id}} P_{n-1} \to 0 \to \dots\right\} & \xrightarrow{\quad g \quad} & X
\end{array}
$$

Then the dotted arrow exists because

$$
\mathrm{Hom}_{C(\mathcal{A})}\left(\left\{0 \longrightarrow P_n \xrightarrow{\text{id}} P_{n-1} \longrightarrow 0\right\}, K\right) = \mathrm{Hom}(P, K_n),
$$

and we have

$$
\begin{array}{ccc}
& & E_n \\
& \nearrow & \downarrow \\
P & \longrightarrow & X_n
\end{array}
$$

since P is projective. We denote $D_n(P) = \left\{0 \longrightarrow P_n \xrightarrow{\text{id}} P_{n-1} \longrightarrow 0\right\}$.

Returning to the general case, we have

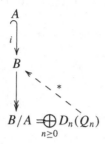

where $Q_n \in \mathcal{P}$. By the special case, we can lift each $D_n(Q_n)$ so that $*$ exists and so

$$B \simeq A \oplus \bigoplus_{n \geq 0} D_n(Q_n).$$

Finally we have

$$
\begin{array}{ccc}
A & \xrightarrow{\ f\ } & E \\
\downarrow & \overset{h}{\nearrow} & \downarrow \\
A \oplus \bigoplus D_n(Q_n) & \longrightarrow & X
\end{array}
$$

where we construct h by lifting each $D_n(Q_n)$ by the special case.   □

Recall that in homotopy theory we have the mapping cylinder construction, namely, given $A \xrightarrow{f} B$, we construct

$$M(f) = (A \times I) \bigcup_{A \times 1} B$$

and have then a factorization of f as follows

$$A \overset{i}{\hookrightarrow} M(f) \xrightarrow{\ p\ } B$$

with i an embedding and p a homotopy equivalence.

The analogue for complexes: let $A. \in C(\mathcal{A})$, and define A_I as follows

$$(A_I)_n = A_n \oplus A_{n-1} \oplus A_n.$$

Notice that if we were working with modules, then we would like to denote an element of $(A_I)_n$ by $(0) \otimes a + (0, 1) \otimes b + (1) \otimes c$, for $a, c \in A_n$ and $b \in A_{n-1}$, and then d is given by

$$d((0) \otimes a + (0,1) \otimes b + (1) \otimes c)$$
$$= (0) \otimes da - (0) \otimes b + (1) \otimes b - (0,1) \otimes db + (1) \otimes dc \,.$$

Back to complexes, the differential $d : (A_I)_n \to (A_I)_{n-1}$ is defined as follows

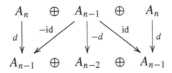

i.e.,
$$d(a,b,c) = (da - b, -db, b + dc) \,.$$

We check

$$d^2(a,b,c) = d(da - b, -db, b + dc)$$
$$= (d(da - b) - (-db), -d(-db), -db + d(b + dc))$$
$$= 0$$

Proposition *To give a map of complexes $A_I \to B$ amounts to giving two maps $f, g : A \to B$ and a homotopy between them, i.e., giving $h : f \to g$ so that $g - f = dh + hd$.*

Proof

$$A_n \oplus A_{n-1} \oplus A_n \longrightarrow B_n$$
$$a \oplus b \oplus c \longmapsto f(a) + h(b) + g(c) \,.$$

Check this is a chain map

$$(a,b,c) \longmapsto f(a) + h(b) + g(c)$$

$$df(a) + dh(b) + dg(c)$$

$$(da - b, -db, b + dc) \longmapsto f(da - b) - hdb + gb + gdc$$

$\square$

Now given $f : A \to B$, a map in $\mathcal{C}(\mathcal{A})$, we define $M(f)$ by push out

$$\begin{array}{ccc} A & \xrightarrow{\;f\;} & B \\ {\scriptstyle(0,\,0,\,a)}\Big\downarrow{\scriptstyle i_1} & & \Big\downarrow \\ A_I & \longrightarrow & M(f) \end{array}$$

Therefore

$$(M(f))_n = A_n \oplus A_{n-1} \oplus B_n$$

and

$$d(a, a', b) = (da - a', -da', fa' + db).$$

We have a canonical factorization

$$A. \xrightarrow{\;i\;} M(f) \xrightarrow{\;p\;} B.,$$

that is,

$$A_n \longrightarrow A_n \oplus A_{n-1} \oplus B_n \longrightarrow B_n$$

given by $i : a \mapsto (a, 0, 0)$ and $p : (a, a', b) \mapsto f(a) + b$, where i is monic, p is epic and

Exercise $p : M(f) \to B$ is a homotopy equivalence.

We can also define A^I so that a map $B. \to A^I$ is the same as a pair $f, g : B. \to A.$ of maps and a homotopy $h : f \to g$. In fact, $(A^I)_n = A_n \times A_{n+1} \times A_n$.

Exercise Find the differential.

Then we can factor $f : B. \to A.$ into $B. \xrightarrow{\;i\;} B. \times_A A^I \xrightarrow{\;p\;} A.$, where i is monic, p is epic and i is homotopy equivalence.

The *mapping cone* of $f : A \to B$ is given by the quotient

$$C(f) = M(f)/\text{Image of } A.$$

Proposition *Let $E. \to X.$ be a quis in $C_+(\mathcal{A})$ and let $P. \in C_+(\mathcal{P})$. Then*

$$[P., E.] \xrightarrow{\;\sim\;} [P., X.]$$

is an isomorphism, where $[\cdot, \cdot]$ denotes homotopy classes.

Proof We factor $E. \to X.$ into $E. \xrightarrow{\;i\;} \widetilde{E}. \xrightarrow{\;p\;} X.$, where p is epic and i is a homotopy equivalence as well as being monic. Since $E. \to X.$ is a quis, it follows that p is also a quis, and moreover,

$$i_* : [P., E.] \simeq [P., \widetilde{E}.],$$

so without loss of generality we assume $E \xrightarrow{\;p\;} X$. is an epic quis.

We must show $[P., E.] \to [P., X.]$ is epic, and to this end, given

we use the lifting theorem.

To show it is monic, suppose $f, f' : P. \to E$. and given a homotopy $h : pf' \to pf$ where $p : E. \to X.$, we have the diagram

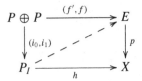

where (i_0, i_1) is monic with cokernel in $C_+(\mathcal{P})$. Use the lifting theorem to conclude that f and f' are homotopic. $\square$

Note that this Proposition gives uniqueness up to canonical homotopy equivalence of projective resolutions... more next time.

Chapter 17
Derived Categories

From last time, we saw

Proposition *If $P. \in C_+(\mathcal{P})$ and $E. \to X.$ is a quis in $C_+(\mathcal{A})$, then*

$$[P., E.] \xrightarrow{\cong} [P., X]$$

is a natural isomorphism.

Remark Check from the proof that this holds without assuming $E.$, $X.$ are in $C_+(\mathcal{A})$ but just in $C(\mathcal{A})$.

Now, recall that we have $K_+(\mathcal{P})$, which has the same objects as $C_+(\mathcal{P})$, but the morphisms are homotopy classes of maps of complexes. We have of course $K_+(\mathcal{P}) \xhookrightarrow{i} K_+(\mathcal{A})$ and claim that i has a right adjoint functor $K_+(\mathcal{P}) \xleftarrow{r} K_+(\mathcal{A})$ defined by choosing for each $A. \in K_+(\mathcal{A})$ a quis $P. \to A.$ with $P. \in C_+(\mathcal{P})$ and setting $r(A.) = P.$.

Claim *that*

$$
\begin{array}{ccc}
\operatorname{Hom}_{K_+(\mathcal{P})}(r(Q.), r(A.)) & \cong & \operatorname{Hom}_{K_+(\mathcal{A})}(iQ., A.) \\
{\scriptstyle =}\Big\uparrow & & {\scriptstyle =}\Big\uparrow \\
[Q., P.] & \xrightarrow{\ \cong\ } & [Q., A.]
\end{array}
$$

via the proposition above.

R. Penner, *Topology and K-Theory*, Lecture Notes in Mathematics 2262,
https://doi.org/10.1007/978-3-030-43996-5_17

These two functors of $A.$ are isomorphic. Thus it follows that r is, in fact, a functor. So we have

$$K_+(\mathcal{P}) \underset{r}{\overset{i}{\rightleftarrows}} K_+(\mathcal{A}) \quad \text{with } ri = \text{id}.$$

For any $A.$, the canonical map $ir(A.) \to A.$ is the quis $P. \to A.$ which we have chosen.

Let $D_+(\mathcal{A})$ be the category with same objects as $C_+(\mathcal{A})$, but with morphisms

$$\operatorname{Hom}_{D_+(\mathcal{A})}(A., B.) = [rA., rB.] \overset{\sim}{\longrightarrow} [rA., B.].$$

Then we notice

Proposition *We have*

$$K_+(\mathcal{A}) \longrightarrow D_+(\mathcal{A})$$
$$[A., B.] \longmapsto [rA., rB.],$$

and this canonical functor is universal with respect to the property that it sends quasi-isomorphisms into isomorphisms

$$
\begin{array}{ccc}
K_+(\mathcal{A}) & \longrightarrow & \mathcal{C} \\
& \searrow \quad \nearrow & \\
& D_+(\mathcal{A}) &
\end{array}
$$

Proof

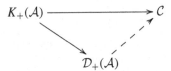

as F carries quasi-isomorphisms to isomorphisms and where the equivalence comes from the fact that the composition gives a clear equivalence of categories.

We want to show that given $F : K_+(\mathcal{A}) \to \mathcal{C}$ inverting the quis, there is a unique way of defining

$$
\begin{array}{ccc}
F_* : \operatorname{Hom}_{D_+(\mathcal{A})}(A., B.) & \dashrightarrow & \operatorname{Hom}(F(A.), F(B.)) \\
\parallel & & \wr\wr \\
[rA., rB.] & \overset{F_*}{\longrightarrow} & \operatorname{Hom}(FrA., FrB.)
\end{array}
$$

for $F(rA.) \xrightarrow{\sim} F(A.)$ and $F(rB.) \xrightarrow{\sim} F(B.)$ since r is quis. Clearly, this will work.   □

$D_+(\mathcal{A})$ is called the *derived category* of complexes in $\mathcal{A}$ bounded below.

$D(\mathcal{A})$ is defined by an analogous universal property from $K(\mathcal{A})$. However, its existence is proved by a process of localization. The idea is to define a map $A. \to B.$ in $D(\mathcal{A})$ to be represented by a diagram

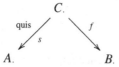

where $C.$ is projective. Think of this as a fraction of s and then proceed as in the construction of localization for a multiplicative system. Consider a projective resolution

$$
\begin{array}{ccc}
P. & \longrightarrow \cdots \longrightarrow P_1 \longrightarrow P_0 \\
\downarrow & \qquad\qquad\qquad \downarrow \\
M[0] & \longrightarrow \cdots \longrightarrow 0 \longrightarrow M \longrightarrow 0.
\end{array}
$$

Let $F : \mathcal{A} \to \mathcal{B}$ be an additive functor between abelian categories. Then given M in $\mathcal{A}$, assuming $\mathcal{A}$ has enough projectives, choose a projective resolution

$$
\cdots \longrightarrow P_2 \longrightarrow P_1 \longrightarrow P_0 \longrightarrow M \longrightarrow 0 \longrightarrow \cdots ,
$$

which is unique up to homotopy. Form

$$
F(P.) : \cdots \longrightarrow F(P_2) \longrightarrow F(P_1) \longrightarrow F(P_0) \longrightarrow 0 \longrightarrow \cdots
$$

and take the homology groups of this chain complex

$$
(L_q F)(M) = H_q(F(P.)) .
$$

Since $P.$ is unique up to homotopy equivalence and F is additive, the homotopy type of $F(P.)$, and hence its homology, depends only on M.

$L_q F$ is called the qth *left-derived functor of* F.

Generalization F induces

$$
\begin{aligned}
C(\mathcal{A}) &\longrightarrow C(\mathcal{B}) , \\
K(\mathcal{A}) &\longrightarrow K(\mathcal{B}) .
\end{aligned}
$$

Namely, given $A. \in K_+(\mathcal{A})$, we have $rA. \in K_+(\mathcal{P})$, and we can define

$$(\mathbb{L}\, F) : K_+\mathcal{A} \to D_+\mathcal{B} \text{ by } (\mathbb{L}\, F)(A.) = F(rA.)\,.$$

This $\mathbb{L}\, F$ has the property that it inverts quasi-isomorphisms. To see this, note that if

then $rA., rA'_.$ are homotopy equivalent complexes, so $\mathbb{L}\, F(A.) \to \mathbb{L}\, F(A'_.)$ is a homotopy equivalence, hence an isomorphism in $D_+(\mathcal{B})$. Thus $\mathbb{L}\, F$ can be viewed as a functor $\mathbb{L}\, F : D_+(\mathcal{A}) \to D_+(\mathcal{B})$.

By definition, if $M \in \text{Ob}\,\mathcal{A}$, then $\mathbb{L}\, F(M[0]) \in D_+(\mathcal{B})$ is an object which is well defined up to canonical isomorphism with

$$H_q(\mathbb{L}\, F(M[0])) = (L_q\, F)(M)\,.$$

Properties of $\{L_q\, F\}$
1: Long exact sequence: Given $0 \to M' \to M \to M'' \to 0$ exact in $\mathcal{A}$, we get a long exact sequence of derived functors

$$\longrightarrow (L_1\, F)\, M'' \longrightarrow (L_0\, F)(M') \longrightarrow (L_0\, F)(M) \longrightarrow (L_0\, F)(M'') \longrightarrow 0\,.$$

2: $(L_q\, F)(P) = 0$ for $q > 0$ if P is projective.
3: There is a canonical map $F \to L.\,F$ which is an isomorphism if and only if F is *right exact*, that is, $0 \to M' \to M \to M'' \to 0$ exact implies that $F(M') \to F(M) \to F(M'') \to 0$ is also exact.

Proof of 2 If P is projective, then a projective resolution is

$$\longrightarrow \underset{2}{0} \longrightarrow \underset{1}{0} \longrightarrow \underset{0}{P} \xrightarrow{\text{id}} P \longrightarrow 0$$

so

$$(L_q\, F)(P) = H_q(F(P[0])) = \begin{cases} F(P) & q = 0, \\ 0 & q \neq 0. \end{cases}$$

Note that Property 3 implies that Property 1 implies that $L_0\, F$ is right exact, for if $\cdots \to P_1 \to P_0 \to M \to 0$ is exact, then $\cdots \to F(P_1) \to F(P_0) \to F(M) \to 0$ is not necessarily exact, but $L_0\, F(M) = H_0(F(P.)) \to F(M)$.

Lemma *The following are equivalent conditions on F*

(a) $0 \to M' \to M \to M'' \to 0$ exact implies $F(M') \to F(M) \to F(M'') \to 0$ exact;

(b) $M' \to M \to M'' \to 0$ exact implies $F(M') \to F(M) \to F(M'') \to 0$ exact.

Proof (b) obviously implies (a). Conversely, given (a) and $N' \xrightarrow{u} N \to N'' \to 0$ exact, break it into

$$0 \to \operatorname{Im} u \to N \to N'' \to 0 \text{ exact,}$$
$$\text{so } F(\operatorname{Im} u) \to F(N) \to F(N'') \to 0 \text{ exact,}$$

and

$$0 \to \operatorname{Ker} u \to N' \to \operatorname{Im} u \to 0 \text{ exact,}$$
$$\text{so } F(\operatorname{Ker} u) \to F(N') \to F(\operatorname{Im} u) \to 0 \text{ exact,}$$

and putting them back together, $F(N') \to F(N) \to F(N'') \to 0$ is also exact.   □

Chapter 18
The First Homotopy Property

Given functors

$$C \underset{g}{\overset{f}{\rightrightarrows}} C'$$

between small categories and a suitable natural transformation $h : f \to g$, we want to show that f and g induce the same map on homology. This is the *first homotopy property*.

Suppose F is a complex of functors $C \to \text{Ab}$. Then we have $H_*(C, F)$, and for $f : C \to C'$ a functor, we have

$$H_*(C, F) = H_*(C', \mathbb{L} f_!(F)).$$

Thus, if we give $\mathbb{L} f_!(F) \xrightarrow{\alpha} F'$, we get an induced homomorphism

$$H_*(C, F) \longrightarrow H_*(C', F')$$

associated to (f, α).

For simplicity we assume that F and F' are just functors (instead of complexes). Then $H_0(\mathbb{L} f_!(F)) = f_!(F)$, so α is the same as a map $f_! F \to F'$ which is in turn the same as $F \to f^* F'$. Thus, given $f : C \to C'$, F, F' and $\alpha : F \to f^* F'$, we get an induced map on homology

$$H_n(C, F) \longrightarrow H_n(C', F'),$$

and in particular, we always have

$$H_n(C, f^* F') \longrightarrow H_n(C', F')$$

© Springer Nature Switzerland AG 2020
R. Penner, *Topology and K-Theory*, Lecture Notes in Mathematics 2262,
https://doi.org/10.1007/978-3-030-43996-5_18

for any $F' : C' \to \mathrm{Ab}$.

Claim *The pair* $f, g : C \to C'$ *together with* $h : f \to g$ *is the same thing as a functor* $C \times I \xrightarrow{H} C'$, *where* $I = \{0 < 1\}$ *is regarded as a category, with* $Hi_0 = f$, *with* $Hi_1 = g$ *and with* $C \underset{i_1}{\overset{i_0}{\rightrightarrows}} C \times I$ *the obvious maps.*

There are the three kinds of maps in $C \times I$:

(i) $(X, 0) \xrightarrow{(u, \mathrm{id})} (X', 0) \longmapsto{\quad H \quad} f(X) \xrightarrow{f(u)} f(X')$,

(ii) $(X, 0) \xrightarrow{(u, <)} (X', 1) \longmapsto{\quad H \quad} f(X)$

$$
\begin{array}{ccc}
 & f(X') & \\
\nearrow {\scriptstyle f(u)} & & \searrow {\scriptstyle h_{X'}} \\
f(X) & & g(X'), \\
\searrow {\scriptstyle h_X} & & \nearrow {\scriptstyle g(u)} \\
 & g(X) &
\end{array}
$$

(iii) $(X, 1) \xrightarrow{(u, \mathrm{id})} (X', 1) \longmapsto{\quad H \quad} f(X) \xrightarrow{f(u)} f(X')$,

where the diamond in (ii) commutes. This justifies the claim. ☐

Consider the projection $p = C \times I \to C$ onto the first factor and choose $F : C \to \mathrm{Ab}$. We compute
$$
H_n(C \times I, p^*F) = H_n(C, \mathbb{L}\, p_!(p^*F)).
$$

But $\mathbb{L}\, p_!(p^*F)$ has the homology groups $L_n\, p_!(p^*F)$ and p is fibered and cofibered so
$$
L_n\, p_!(p^*F)(X) = H_n(p^{-1}X, p^*F|_{p^{-1}(X)})
$$
$$
= H_n(I, F(X) \text{ as a constant functor}).
$$

(∗) Now, I has a final object e, so $\varinjlim_I F = F(e)$, whence this is an exact functor of F, and so $L_q \varinjlim_I (\cdot) = H_q(I, \cdot) = 0$, for all $q > 0$. It follows that

$$L_n \, p_!(p^*F)(X) = \begin{cases} 0, & n \neq 0, \\ F(X), & n = 0, \end{cases}$$

so $\mathbb{L} \, p_!(p^*F)$ is quis to F. Therefore $H_n(\mathcal{C} \times I, p^*F) = H_n(\mathcal{C}, F)$.

Exercise 1 Compute $H_n(\mathcal{C} \times I, G)$ for any G on $\mathcal{C} \times I$. One of the embeddings i_0, i_1 has p as adjoint of the sort that (by last time) $\mathbb{L} \, p_!$ is i_j^*. Quillen is not sure that this will work.

Proposition (First Homotopy Property) *Suppose given* $\mathcal{C} \underset{g}{\overset{f}{\rightrightarrows}} \mathcal{C}'$ *with* $h : f \to g$ *and* $F' : \mathcal{C}' \to \mathrm{Ab}$, *such that* h *induces an isomorphism* $f^*F' \to g^*F'$. *Recall*

$$(f^*F')(X) = F'(f(X)) \xrightarrow{F'(h_X)} F'(g(X)) = (g^*F')(X),$$

and we claim

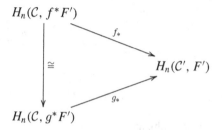

commutes, where f_* *here means maps induced on homology, not the adjoint of* f^* *and likewise for* g_*.

Example If F' is the constant functor with value $A \in \mathrm{Ob} \, \mathrm{Ab}$, then the proposition says that the following diagram commutes.

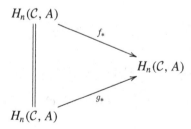

Proof of Proposition We have

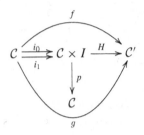

and given F', we know $F'(fX) \xrightarrow[F'(h_X)]{\sim} F'(gX)$. But $H^*(F')$ is constant on the fibers of p, and so

$$H^*(F') \cong p^* f^*(F') \overset{d}{=} p^* G .$$

Thus we have the commutative diagram

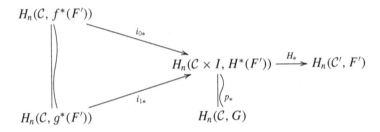

where i_{0*}, i_{1*} and p_* are isomorphisms.   □

We noted in (∗) that if C has a final object, then $\varinjlim C$ is exact so $H_n(C, F) = 0$, for all $n \neq 0$. However, the conclusion is **not true** if C has just an initial object.

Example

$$a \underset{c}{\overset{b}{\rightrightarrows}} \equiv C \rightsquigarrow F(a) \overset{F(b)}{\underset{F(c)}{\rightrightarrows}}$$

and $H_n(C)$ fits into a Mayer–Vietoris type sequence

$$0 \longrightarrow H_1(C, F) \longrightarrow F(a) \longrightarrow F(b) \oplus F(c) \longrightarrow H_0(C, F) \longrightarrow 0,$$

with all $H_n(C, F) = 0$ for $n \neq 0, 1$.
    This example was abandoned apologetically.

    We know, for $i_b :$ point $\to C$ with image b, that

$$H_n(\mathcal{C}, i_{b!}A) = H_n(\mathcal{C}, \mathbb{L}\, i_{b!}(A)) \text{ since } i_{b!} \text{ is exact}$$
$$= H_n(\text{point}, A)$$
$$= \begin{cases} A, & n = 0, \\ 0, & \text{else.} \end{cases}$$

Claim *For a constant functor A and a category $\mathcal{C}$ with an initial object that*

$$H_n(\mathcal{C}, A) = \begin{cases} A, & n = 0, \\ 0, & \text{else,} \end{cases}$$

using the homotopy property, i.e., that if $\mathcal{C} \underset{g}{\overset{f}{\rightrightarrows}} \mathcal{C}'$ and $h : f \to g$, then

$H_n(\mathcal{C}, A) \underset{g_*}{\overset{f_*}{\rightrightarrows}} H_n(\mathcal{C}', A)$ *coincide.*

Thus, if we have a pair of adjoint functors $\mathcal{C} \underset{g}{\overset{f}{\rightleftarrows}} \mathcal{C}'$, then

$$H_n(\mathcal{C}, A) \underset{g_*}{\overset{f_*}{\rightleftarrows}} H_n(\mathcal{C}', A) \ ,$$

and so f_* and g_* are inverses.

Proof point $\overset{i_X}{\longrightarrow} \mathcal{C}$ has a right adjoint $Y \mapsto *$, and $\text{Hom}_{\mathcal{C}}(i_X *, Y) = \text{Hom}_{\text{point}}(*, *)$, when X is initial in $\mathcal{C}$. Thus $H_n(\text{point}, A) = H_n(\mathcal{C}, A)$ when $\mathcal{C}$ has an initial object. $\qquad\qquad\square$

Claim *We have $H_n(\mathcal{C}, A) \cong H_n(\mathcal{C}^{\text{op}}, A)$.*

Proof We form over $\mathcal{C}^{\text{op}} \times \mathcal{C}$ the cofibered category $\mathcal{E}$ belonging to the functor $(X, Y) \mapsto \text{Hom}_{\mathcal{C}}(X, Y)$. The objects of $\mathcal{E}$ are arrows $X \overset{f}{\longrightarrow} Y$, and a morphism is a diagram

$$
\begin{array}{ccc}
X & \overset{f}{\longrightarrow} & Y \\
\big\uparrow{\scriptstyle \text{in } \mathcal{C}^{\text{op}}} & & \big\downarrow{\scriptstyle \text{in } \mathcal{C}} \\
X' & \underset{f'}{\longrightarrow} & Y'
\end{array}
$$

We will show that the two obvious functors

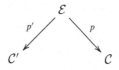

induce isomorphisms on homology.

To this end, the functor

$$p : \mathcal{E} \longrightarrow \mathcal{C}$$
$$: (X \xrightarrow{f} Y) \longmapsto Y$$

is the composite of cofibered hence cofibered (exercise) with

$$p^{-1} Y = (\mathcal{C}/Y)^{\mathrm{op}}.$$

But $\mathcal{C}/Y$ has a final object, which implies that $(\mathcal{C}/Y)^{\mathrm{op}}$ has an initial one.

Similarly,

$$p' : \mathcal{E} \longrightarrow \mathcal{C}'$$
$$: (X \xrightarrow{f} Y) \longmapsto X$$

is cofibered with $(p')^{-1}(X) = X \backslash \mathcal{C}$, which also has an initial object. Now

$$(\dagger) \qquad\qquad\qquad H_n(\mathcal{E}, A) = H_n(\mathcal{C}, \mathbb{L} \, p_!(A))$$

and $\mathbb{L} \, p_!(A)$ has homology groups given by

$$L_q \, p_!(A)(Y) = H_q(p^{-1} Y, A) \quad \text{since } p \text{ cofibered}$$
$$= \begin{cases} A, \; q = 0, \\ 0, \; \text{else}, \end{cases} \quad \text{since } p^{-1} Y \text{ has an initial object.}$$

Therefore $\mathbb{L} \, p_! A = A$, so $(\dagger)$ follows, and similarly for p'. $\qquad\qquad\qquad\square$

Chapter 19
Group Completions and Grothendieck Groups

We want to define the **group completion** of an abelian monoid S, (being a monoid means that S has a commutative and associative operation $+$ and the unit 0 exists). More precisely:

Problem Construct an abelian group G and a monoid morphism $S \xrightarrow{u} G$ which is universal. G is then called the *group completion of S*.

Solution Take $G = (S \times S)/\sim$ where $\sim$ is generated by $(s_1, s_2) \sim (s + s_1, s + s_2)$. Check this works and yields an abelian group with $-(s_1, s_2) = (s_2, s_1)$ and $u : s \mapsto (s, 0)$, so $(s_1, s_2) = u(s_1) - u(s_2)$. This is Grothendieck's construction.

Example $S = \mathbb{N}$.

$$\mathbb{N} \times \mathbb{N} \longrightarrow \mathbb{Z}$$
$$(n_1, n_2) \longmapsto n_1 - n_2 .$$

We get $G = (\mathbb{N} \times \mathbb{N})/\sim \to \mathbb{Z}$, and this map is clearly a bijection.

More generally, if S is a sub-monoid of an abelian group A so that $A = S + (-S)$, then $A \approx$ group completion of S. Notice that we can do this for a category leading to a groupoid, but this loses most structure, e.g., higher homotopy.

If X is a set on which S acts, construct $S^{-1} X = (X \times S)/\sim$ where $\sim$ is generated by $(x, s) \sim (s'x, s' + s)$. $S^{-1}X$ is a set with S action $s'(x, s) = (s'x, s)$, and each s' acts invertibly on $S^{-1}X$. We also have a map $X \to S^{-1}X$ which is universal for maps from X to S-sets on which S acts invertibly.

Example Localization for S a multiplicative system in a commutative ring A and X an A-module.

Note that the group completion of S is $S^{-1} S$.

© Springer Nature Switzerland AG 2020

R. Penner, *Topology and K-Theory*, Lecture Notes in Mathematics 2262,
https://doi.org/10.1007/978-3-030-43996-5_19

Exercise Calculate $S^{-1}S$ where $S = \mathbb{Z}$ under multiplication.

Let A be a ring and $\mathcal{P}_A$ the category of finitely generated projective A-modules P, so Hom $(P, \cdot)$ is exact and in particular id : $P \to P$ comes from $P \to A^n$. Take S to be the set of isomorphism classes of $\mathcal{P}_A$ and addition on S induced by $(P, Q) \mapsto P \oplus Q$.

The *Grothendieck group* of $\mathcal{P}_A$ is the group completion of S, denoted $K_0 A$.

Example 1 If F is a field, then isomorphism classes of $\mathcal{P}_F \cong \mathbb{N}$ by dimension whence $K_0 F = \mathbb{Z}$. More generally, if every $P \in \mathcal{P}_A$ is free, then $K_0 A \cong \mathbb{Z}$. For instance

(1) local commutative rings by Nakayama,
(2) principal ideal domains,
(3) [1976] (Serre problem) $A = k[X_1, \ldots, X_n]$ for k a principal ideal domain.

Example 2 Consider a Dedekind domain A, e.g., algebraic integers in an algebraic number field. Any finitely generated torsion free A-module is projective and is a direct sum $P = \mathcal{A}_1 \oplus \ldots \oplus \mathcal{A}_n$, where the $\mathcal{A}_i$ are ideals. Actually $P = A^{n-1} \oplus \mathcal{A}$ where $\mathcal{A}$ is an ideal. In fact, the exterior algebra $\Lambda^n P \cong \Lambda^{n-1} A^{n-1} \otimes \Lambda^1 \mathcal{A} \cong \mathcal{A}$, and the class of $\mathcal{A}$ is the ideal class group Pic A, an invariant of P. Moreover, it depends additively on the projective module. The isomorphism classes in Pic A are therefore $S \subset \mathbb{N} \times$ Pic A in the obvious way and $K_0 A = \mathbb{Z} \oplus$ Pic A.

Example 3 [Serre]: Consider a compact Hausdorff space X and let A be the ring of continuous $\mathbb{C}$-valued functions on X. Serre and Swan showed there is an equivalence between $\mathcal{P}_A$ and the category of complex vector bundles E over X as follows. If E is a vector bundle, let $\Gamma(X, E)$ denote the continuous global sections of E. $\Gamma(X, E)$ is a finitely generated projective A-module: Any E is a direct summand of a trivial vector bundle, and we get a surjection $X \times \mathbb{C}^n \twoheadrightarrow E$ which splits. Thus $\Gamma(X, E)$ is a direct summand of $\Gamma(X, X \times \mathbb{C}^n) = A^n$. Conversely, given $P \in \mathcal{P}_A$, express P as a summand of A^n. This gives an idempotent $n \times n$ matrix M over A. Then M can be viewed as an endomorphism of $X \times \mathbb{C}^n$ which splits the bundle by idempotence. To see locally triviality, note that Im $M = \text{Ker}\{1 - M\}$ again by idempotence.

Now, from algebraic topology for X connected, we have isomorphism classes of n-dimensional vector bundles over X corresponding to $[X, BU_n]$.

Instead of $\mathcal{P}_A$, we can use any category $\mathcal{C}$ having a set of isomorphism classes together with a functor $\mathcal{C} \times \mathcal{C} \xrightarrow{\perp} \mathcal{C}$, where $X \times Y \longmapsto X \perp Y$, so that

$$X \perp Y \simeq Y \perp X,$$
$$(X \perp Y) \perp Z \simeq X \perp (Y \perp Z),$$
there is $0 \in \mathcal{C}$ so that $0 \perp X \simeq X$.

Given all this, we can group complete the isomorphism classes in $\mathcal{C}$ to get a Grothendieck group $K_0 \mathcal{C}$.

Exercise For G a finite group, $\mathcal{C}$ the category of finite G-sets and $X \perp Y$ the disjoint union of G-sets, describe K_0, called the *Burnside ring of G*.

Grothendieck's original: Let $\mathcal{M}$ be a full subcategory of an abelian category $\mathcal{A}$. Assume $0 \in \mathcal{M}$ and $\mathcal{M}$ is closed under extensions, i.e., if $0 \to M' \to A \to M \to 0$ is exact with $M, M' \in \mathcal{M}$, then $A \in \mathcal{M}$ as well.

The *Grothendieck group* $K_0 \mathcal{M}$ is an abelian group together with a map

$$M \longmapsto [M]$$

from isomorphism classes of $\mathcal{M}$ to $K_0 \mathcal{M}$ so that for any exact sequence $0 \to M' \to M \to M'' \to 0$, we have $[M] = [M'] + [M'']$, and moreover, this map is universal.

Remark $K_0(\mathcal{M}, \oplus) \twoheadrightarrow K_0 \mathcal{M}$.

Example For $\mathcal{M}$ the category of finite abelian groups, we get

$$K_0(\mathcal{M}, \oplus) = \bigoplus_{\substack{\text{primes } p \\ r \in \mathbb{Z}_+}} \mathbb{Z},$$

$$K_0(\mathcal{M}) = \bigoplus_{\text{primes } p} \mathbb{Z}.$$

Chapter 20
Devissage and Resolution Theorems

Suppose that $\mathcal{M}$ is a full subcategory of an abelian category $\mathcal{A}$ where $\mathcal{M}$ is closed under finite direct sums, contains 0 and whose isomorphism classes form a set. Recall that $K_0\,\mathcal{M}$ is an abelian group together with a universal map

$$\text{Ob}\,\mathcal{M} \longrightarrow K_0\,\mathcal{M}$$

so that

$$0 \longrightarrow M' \longrightarrow M \longrightarrow M'' \longrightarrow 0$$

exact in $\mathcal{M}$ implies that $[M] = [M'] + [M'']$.

Remark If $\mathcal{M} = \mathcal{P}_A$, then exact sequences split, so $K_0(\mathcal{M}, \oplus) = K_0\,\mathcal{M}$. In general, $K_0\,\mathcal{M}$ is a quotient of $K_0(\mathcal{M}, \oplus)$.

Example Let $\mathcal{M}$ be the category of finite abelian p-groups. Recall that Krull–Schmidt says every $M \in \mathcal{M}$ is a direct sum of indecomposables in a unique way up to isomorphism. Thus isomorphism classes of $\mathcal{M}$ are given by the free abelian monoid generated by indecomposable isomorphism classes, i.e., $\mathbb{Z}/p^r$, $r = 1, \ldots$. Thus $K_0(\mathcal{M}, \oplus)$ is the free abelian group generated by

$$\mathbb{Z}/p^r\,\mathbb{Z}, \quad r \geq 1.$$

Whilst in general if we have a filtration

$$0 = F_{-1}\,M \subset F_0\,M \subset \ldots \subset F_n\,M = M$$

of $M \in \mathcal{M}$ so that each $F_p\,M$ and $F_p\,M / F_{p-1}\,M \in \mathcal{M}$, then in $K_0\,\mathcal{M}$, we have

$$(\dagger)\ M = \sum_{p=0}^{n} [F_p\,M / F_{p-1}\,M].$$

© Springer Nature Switzerland AG 2020
R. Penner, *Topology and K-Theory*, Lecture Notes in Mathematics 2262,
https://doi.org/10.1007/978-3-030-43996-5_20

Thus in our example, every M has a filtration with quotients $\mathbb{Z}/p\mathbb{Z}$ so $K_0 M = \mathbb{Z}$. Namely we have the map assigning the integral filtration length which is a homomorphism $K_0 M \to \mathbb{Z}$ by universality.

More generally, suppose $\mathcal{M}$ is an abelian category where every object has the descending chain condition and the ascending chain condition on its sub-objects. Then Jordan–Hölder applies, so we can speak of the multiplicity $m_\sigma(M)$ of the simple object σ in any composition series for M. We have $M \longmapsto \{m_\sigma(M)\}$ in the free monoid generated by simple objects and get a map $K_0 \mathcal{M} \xrightarrow{\approx} \bigoplus_\sigma \mathbb{Z}$.

Devissage Theorem *(Grothendieck) Let $\mathcal{A}$ be an abelian category and $\mathcal{A}' \subset \mathcal{A}$ full so that also $\mathcal{A}'$ is abelian. Assume every A in $\mathcal{A}$ has a finite filtration*

$$0 = F_{-1} \subset \cdots \subset F_n = A$$

so that $F_p/F_{p-1} \in \mathcal{A}'$. Then $K_0 \mathcal{A}' \xrightarrow{\sim} K_0 \mathcal{A}$.

Example 1 Take $\mathcal{A}$ to be finite abelian p-groups and $\mathcal{A}'$ to be finitely generated $\mathbb{Z}/p$-modules.

Example 2 Suppose A is a noetherian ring, $\mathcal{M} = \text{Modf}(A)$, the finitely generated A-modules, and $I \subset A$ is a nilpotent ideal. Take $\mathcal{M}' = \text{Modf}(A/I)$ where

$$0 = I^n M \subset \ldots \subset IM \subset M.$$

Thus $K_0 \text{Mod} f(A/I) = K_0 \text{Modf}(A)$.

The proof of the Devissage Theorem is based on the Schreier refinement lemma.

Lemma *(Schreier) Suppose M has two filtrations $\cdots \subset F_p M \subset F_{p+1} M \subset \cdots$ and $\cdots \subset F'_q M \subset F'_{q+1} M \subset \cdots$, neither necessarily terminating. Then F' induces a filtration on $\text{gr}^F M = \bigoplus_p F_p M/F_{p-1} M$ via $F'_q(F_p M) = F'_q M \cap F_p M$, i.e.,*

$$F'_q(F_p M/F_{p-1} M) = \text{Image}\{F'_q \cap F_p \longrightarrow F_p/F_{p-1}\}.$$

Similarly F induces a filtration on $\text{gr}^{F'} M = \bigoplus F'_q/F'_{q-1}$. Then we have

$$\text{gr}^{F'}(\text{gr}^F(M)) = \text{gr}^F(\text{gr}^{F'}(M)).$$

Proof We have the formulae

$$F'_q(F_p/F_{p-1}) = F'_q \cap F_p + F_{p-1}/F_{p-1},$$

$$\text{gr}^{F'}_q(F_p/F_{p-1}) = \frac{F'_q \cap F_p + F_{p-1}}{F'_{q-1} \cap F_p + F_{p-1}} = \frac{F'_q \cap F_p}{(F'_{q-1} \cap F_p + F_{p-1}) \cap (F'_q \cap F_p)}$$

but

$$((F'_{q-1} \cap F_p) + F_{p-1}) \cap (F'_q \cap F_p) = F'_{q-1} \cap F_p + F_{p-1} \cap F'_q.$$

$\square$

Remark The refinement lemma fails for 3 filtrations since the modular law fails.

Proof of Devissage Given $M \in \mathcal{A}$, choose a filtration $F_p M$ exhausting M with quotients $\mathrm{gr}^F_p(M) \in \mathcal{A}'$ and consider

$$\sum_p [\mathrm{gr}^F_p M] \in K_0 \mathcal{A}'.$$

Claim *This does not depend on the choice of* $\{F_p M\}$. *This follows easily from the previous lemma and* (†).

Thus we get a map $\gamma : \mathrm{Ob}\,\mathcal{A} \to K_0 \mathcal{A}'$. Then $0 \to M' \to M \to M'' \to 0$ exact in $\mathcal{A}$ implies that $\gamma(M) = \gamma(M') + \gamma(M'')$. By universality, we get $\gamma : K_0 \mathcal{A} \to K_0 \mathcal{A}'$, which is inverse to the map induced by inclusion. $\square$

Example Take $\mathcal{A} = \mathrm{Modf}(\mathbb{Z})$, so $\mathcal{P}_\mathbb{Z}$ is the category of free finitely generated abelian groups, and we have $K_0 \mathcal{P}_\mathbb{Z}) = \mathbb{Z}$. Now, any finitely generated abelian group has a resolution

$$0 \longrightarrow \mathbb{Z}^q \longrightarrow \mathbb{Z}^p \longrightarrow M \longrightarrow 0$$

so

$$[M] = [\mathbb{Z}^p] - [\mathbb{Z}^q]$$
$$= (p - q)[\mathbb{Z}],$$

whence $K_0 \mathcal{P}_\mathbb{Z} \twoheadrightarrow K_0(\mathrm{Mod}\,f(\mathbb{Z}))$ is in fact an isomorphism.

Resolution Theorem *(Grothendieck) Suppose A is a noetherian ring so that every finitely generated A-module M has a finite resolution of projectives*

$$(*) \qquad\qquad 0 \longrightarrow P_n \longrightarrow \ldots \longrightarrow P_0 \longrightarrow M \longrightarrow 0$$

A is said to be regular *in this case.*

Then

$$K_0(\mathcal{P}_A) \xrightarrow{\sim} K_0 \mathrm{Modf}(A).$$

The idea is to define

$$K_0 \mathrm{Modf}(A) \longrightarrow K_0 \mathcal{P}_A$$
$$[M] \longmapsto \sum_{i=0}^n (-1)^i [P_i].$$

If this is well defined, then it is an inverse to $K_0 \mathcal{P}_A \to K_0 \operatorname{Modf}(A)$. To see this, decompose (∗) into

$$0 \longrightarrow R_1 \longrightarrow P_0 \longrightarrow M \longrightarrow 0$$

$$0 \longrightarrow R_2 \longrightarrow P_1 \longrightarrow R_1 \longrightarrow 0$$

$$\vdots$$

$$0 \longrightarrow 0 \longrightarrow P_n \longrightarrow R_n \longrightarrow 0.$$

Thus in $K_0 \operatorname{Modf}(A)$, we have

$$[M] = \sum_{i=0}^{n} (-1)^i [P_i],$$

and it remains as an exercise only to show it is well defined.   □

Chapter 21
Exact Sequences of Homotopy Classes

Suppose that $F : \mathcal{A} \to \mathcal{B}$ is an additive functor between abelian categories and $\mathcal{A}$ has enough projectives. For M in $\mathcal{A}$, choose a projective resolution, i.e., a quis $P_{\cdot} \to M[0]$. Then $F(P_{\cdot}) = \mathbb{L}F(M[0])$ is unique up to canonical isomorphism in $D_+(\mathcal{B})$, and we define

$$L_i F(M) = H_i \, \mathbb{L}F(M[0])$$
$$= H_i \, F(P_{\cdot}).$$

Claim *Given* $0 \to M' \to M \to M'' \to 0$ *exact in* $\mathcal{A}$, *then we have a natural long exact sequence*

(†) $\quad \cdots \longrightarrow L_1 F M' \longrightarrow L_1 F M \longrightarrow L_1 F M'' \longrightarrow L_0 F M' \longrightarrow L_0 F M \longrightarrow L_0 F M'' \longrightarrow 0.$
We shall prove this more generally for $M_{\cdot}$ *in* $C_+(\mathcal{A})$ *and* $P_{\cdot} \to M_{\cdot}$, *and then the sequence* (†) *continues on to the right:* $L_0 F M'' \to L_{-1} F M' \to \cdots$.

Proof Start with projective resolutions of M' and M

(∗)

and note that the induced $P' \to P$ comes from $[P', P] \xrightarrow{\sim} [P', M]$.

Thus we get $f : P' \to P$ and a homotopy $h : ip' \to pf$, i.e., $dh + hd = pf - ip'$.

© Springer Nature Switzerland AG 2020
R. Penner, *Topology and K-Theory*, Lecture Notes in Mathematics 2262,
https://doi.org/10.1007/978-3-030-43996-5_21

Note that given

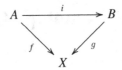

with a homotopy h from f to g, we can form $A \xrightarrow{i_0} A \times I \cup_i B$ and get a map $M(i) \to X$, where $M(\bullet)$ is the mapping cylinder of $\bullet$. satisfying

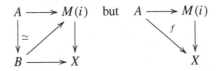

where the right-hand diagram and the bottom triangle of the left-hand diagram each commutes and the diagonal morphism of the left-hand diagram is a homotopy equivalence.

Recall we have $M(f : P' \to P) = P'_I \coprod_{P'} P$ as $M(f : P' \to P)_n = P'_n \oplus P'_{n-1} \oplus P_n$ with the appropriate differential.

Thus we can replace P in $(*)$ by $M(f : P' \to P)$ to get

$$
\begin{array}{ccc}
P'_n & \xrightarrow{(\mathrm{id},0,0)} & P'_n \oplus P'_{n-1} \oplus P_n \\
\downarrow{\scriptstyle p'} & & \downarrow{\scriptstyle (ip',\pm h,p)} \\
M'_n & \xrightarrow{\quad i \quad} & M_n
\end{array}
$$

where we can check that with an appropriate choice of sign in $\pm h$ that (ip', h, p) is a chain map.

We get

$$(**) \qquad
\begin{array}{ccccccccc}
0 & \longrightarrow & P' & \hookrightarrow & M\left(P' \xrightarrow{f} P\right) & \longrightarrow & \mathrm{Cone}\,(f) & \longrightarrow & 0 \\
& & \downarrow{\scriptstyle \mathrm{quis}} & & \downarrow{\scriptstyle \mathrm{quis}} & & \downarrow{\scriptstyle \mathrm{quis}} & & \\
0 & \longrightarrow & M' & \longrightarrow & M & \longrightarrow & M'' & \longrightarrow & 0
\end{array}
$$

where $\mathrm{Cone}(\bullet)$ is the mapping cone of $\bullet$ and the leftmost square commutes, proving the rightmost map to be a quis by a long exact and the 5-lemma.

It is also clear that $\mathrm{Cone}\,(f)$ is projective.

Now apply F to get a sequence

($\ddagger$) $0 \longrightarrow F(P') \longrightarrow FM(f) \longrightarrow F \text{ Cone}(f) \longrightarrow 0,$

which is exact because the sequence (∗∗) splits since $\text{Cone}(f) \in C_+(\mathcal{P})$. In fact $F \text{ Cone}(f) = \text{Cone } Ff$ for F additive.

Now take the long exact homology sequence of ($\ddagger$). Naturality follows easily.   □

Suppose given a module exact sequence and two projective resolutions, so we have

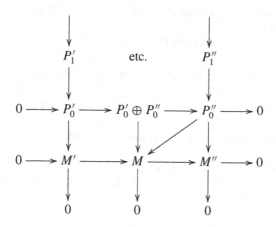

This does not work for complexes because projective in each degree does not imply projective.

Exercise Describe projective objects in $C_+(\mathcal{A})$; the answer is **not** all of $C_+(\mathcal{P})$.

Puppe Sequences Fix an additive category $\mathcal{A}$ and suppose we have a map $f : A \to B$ of complexes. Then we get, for any complex N, a long exact sequence

$$\cdots \longleftarrow [S^{-1}\text{Cone} f, N] \longleftarrow [A, N] \longleftarrow [B, N] \longleftarrow [\text{Cone} f, N] \longleftarrow$$

$$\longleftarrow [SA, N] \longleftarrow [SB, N] \longleftarrow [S \text{ Cone} f, N] \longleftarrow \cdots$$

where $SA = \text{Cone } A \to 0$ so $(SA)_n = A_{n-1}$ and $d : (SA)_n \to (SA)_{n-1}$ is given by $-d : A_{n-1} \to A_{n-2}$.

Lemma *Suppose* $0 \to A' \to A \to A'' \to 0$ *is an exact sequence of complexes which splits in each dimension. Then for any* N, *we have exact sequences*

$$[A', N] \longleftarrow [A, N] \longleftarrow [A'', N],$$

$$[N, A'] \longrightarrow [N, A] \longrightarrow [N, A''].$$

Proof We introduce the **complex** Hom.(A, N), where

$$\operatorname{Hom}_p(A, N) = \prod_n \operatorname{Hom}(A_n, N_{n+p})$$

with $df = d \circ f - (-1)^p f \circ d$.

Remark 0-cycles in Hom.(A, N) are $f = (f : A_n \to N_n)$ so that $d \circ f = f \circ d$, i.e., maps of complexes $A \to N$. A zero-boundary is a map $f = (f : A_n \to N_n)$ of the form $f = d \circ h + h \circ d$ for some $h = (h : A_n \to N_{n+1})$ i.e., those maps homotopic to zero. Thus

$$H_0 \operatorname{Hom.}(A, N) = [A, N].$$

Now, if $0 \to A' \to A \to A'' \to 0$ splits in each dimension, then

$$0 \longrightarrow \operatorname{Hom.}(N, A') \longrightarrow \operatorname{Hom.}(N, A) \longrightarrow \operatorname{Hom.}(N, A'') \longrightarrow 0$$

is exact as a product of exact sequences. Take the homology. □

Remark For complexes of abelian groups for example, we have $K. \otimes L.$ with $d(x \otimes y) = dx \otimes y + (-1)^{|x|} x \otimes dy$. Then the differential in Hom.(A, N) is the one making the evaluation map

$$\operatorname{Hom.}(A., N.) \otimes N. \longrightarrow N.$$

a map of complexes.

Chapter 22
Spectral Sequences

Recall the

Lemma *Suppose* $0 \to A' \to A \to A'' \to 0$ *is exact in* $C(\mathcal{A})$ *and splits in each dimension. Then*

$$[A', X] \longleftarrow [A, X] \longleftarrow [A'', X]$$

is exact.

Recall also that given $A \xrightarrow{f} B$, *we can construct* $M(f) = A_I \coprod_A B$. *Thus, we have the commutative diagram*

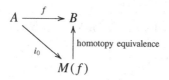

and set $C(f) = \mathrm{Cok}\, i_0$. *The Lemma applies to*

$$0 \longrightarrow A \xrightarrow{i_0} M(f) \longrightarrow C(f) \longrightarrow 0$$

because in dimension n we have

$$0 \longrightarrow A_n \longrightarrow A_n \oplus A_{n-1} \oplus B_n \longrightarrow A_{n-1} \oplus B_n \longrightarrow 0,$$

and so we get an exact

© Springer Nature Switzerland AG 2020

R. Penner, *Topology and K-Theory*, Lecture Notes in Mathematics 2262,

https://doi.org/10.1007/978-3-030-43996-5_22

$$[A, X] \longleftarrow [M(f), X] \longleftarrow [C(f), X]$$

$$[B, X]$$

with f^*, j^* and the middle vertical map to $[B, X]$.

where $j : B \hookrightarrow C(f)$.
  Now construct

$$A \xrightarrow{\ f\ } B \xrightarrow{\ j\ } C(f) \xrightarrow{\ j'\ } C(j) \longrightarrow C(j') \longrightarrow \cdots$$

and get exact sequences of homotopy classes

$$[A, X] \longleftarrow [B, X] \longleftarrow [C(f), X] \longleftarrow [C(j), X] \longleftarrow [C(j'), X].$$

Note that in general $C(f)/B = S(A)$ and $C(j)/C(f) = S(B)$.

(†) **Lemma:** If $A \xrightarrow{\ f\ } B$ is injective and splits in each dimension, then the canonical map

$$C(f) \longrightarrow B/A$$

is a homotopy equivalence.

Thus $C(f)/B \xleftarrow[\text{equivalence}]{\text{homotopy}} C(j)$ and $C(j)/C(f) \xleftarrow[\text{equivalence}]{\text{homotopy}} C(j')$,
$\quad\quad\;\; \| \atop {\scriptstyle S(A)} \quad\quad\quad\quad\quad\quad\quad\quad\quad\quad\;\; \| \atop {\scriptstyle S(B)}$

and so we get the exact sequence

$$\cdots \longleftarrow [A, X] \longleftarrow [B, X] \longleftarrow [C(f), X] \longleftarrow [SA, X] \longleftarrow [SB, X] \longleftarrow \cdots$$

Point of Lemma (†): We have

$$
\begin{array}{ccccccccc}
0 & \longrightarrow & A & \xrightarrow{\ f\ } & B & \longrightarrow & B/A & \longrightarrow & 0 \\
  &   & \| &   & \uparrow &   & \uparrow &   & \\
0 & \longrightarrow & A & \longrightarrow & M(f) & \longrightarrow & C(f) & \longrightarrow & 0
\end{array}
$$

The first lemma gives the following exact

$$
\begin{array}{ccccccccccc}
[A, X] & \longleftarrow & [B, X] & \longleftarrow & [B/A, X] & \longleftarrow & H_1 \,\text{Hom.}(A, X) & \longleftarrow & \bullet \\
\| & & \wr \downarrow & & & & \| & & \wr \downarrow \\
[A, X] & \longleftarrow & [M(f), X] & \longleftarrow & [C(f), X] & \longleftarrow & H_1 \,\text{Hom.}(A, X) & \longleftarrow & H_1 \text{Hom.} C(f)
\end{array}
$$

so the 5-lemma implies $[B/A, X] \xrightarrow{\ \sim\ } [C(f), X]$ and by Yoneda $C(f) \to B/A$ is an isomorphism in $K\,\mathcal{A}$, as desired. $\square$

Spectral Sequences

Situation: Suppose K is a complex provided with an increasing filtration by sub-complexes

$$0 \subset \cdots \subset F_{p-1}K \subset F_p K \subset F_{p+1}K \subset \cdots \subset K$$

and suppose $T_* = \{T_n\}$ is a "∂-functor on complexes", that is, it associates **naturally** to a short exact sequence of complexes $0 \to K' \to K \to K'' \to 0$ a long exact sequence

$$\cdots \longrightarrow T_{n+1}K'' \overset{\partial}{\longrightarrow} T_n K' \longrightarrow T_n K \longrightarrow T_n K'' \overset{\partial}{\longrightarrow} T_{n-1}K' \longrightarrow \cdots$$

in an abelian category $\mathcal{B}$.

Problem To relate $T_* K$ with $T_* \operatorname{gr}_* K$ where

$$\operatorname{gr}_p K = F_p K / F_{p-1} K .$$

Useful diagram: Given

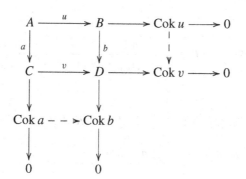

where the top-left square commutes, then we have

$$\operatorname{Cok} v / \operatorname{Cok} u$$

$$\|$$

$$\operatorname{Cok} b / \operatorname{Cok} a = D/(\operatorname{Im} b \oplus \operatorname{Im} v) .$$

Proof Diagram chase.   □

Consider the induced filtration on $T_*(K)$ and let $F_p T_* K = \operatorname{Im}\{T_* F_p K \to T_* K\}$.
  Can we see $F_p T K / F_{p-1} T K$ inside $T K_p / K_{p-1}$?
  Use the exact sequence

$$T_{n+1}(K/F_p K) \overset{\partial}{\longrightarrow} T_n F_p K \longrightarrow T_n K \longrightarrow T_n K/F_p K$$

to get

$$\mathrm{Cok}\{T_{n+1}K/F_p\,K \xrightarrow{\;\partial\;} T_n\,F_p\,K\} = F_p\,T\,K\,.$$

Then we find

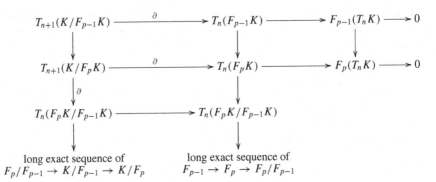

$$\text{long exact sequence of} \qquad \text{long exact sequence of}$$
$$F_p/F_{p-1} \to K/F_{p-1} \to K/F_p \qquad F_{p-1} \to F_p \to F_p/F_{p-1}$$

Thus by the useful diagram, $F_p\,T_n\,K/F_{p-1}\,T_n\,K$ is isomorphic to a sub-quotient of $T_n\,F_p\,K/F_{p-1}K.$

Precisely, if we put

$$F_p\,T_n\,K/F_{p-1}\,T_n\,K \cong \frac{\mathrm{Im}\{T_nF_pK \to T_n\,F_p/F_{p-1}\}}{\mathrm{Im}\{T_{n+1}K/F_pK \to T_nF_p/F_{p-1}\}}\,,$$

then we derive

Corollary *If the filtration is finite and exhausts K and if $T_n(F_p/F_{p-1}) = 0$, then $T_n\,K = 0$.*

Define $T_n\,F_p\,K/F_q\,K = T_n(p,q)$ and set

$$Z^r_{p,q} = \mathrm{Im}\{T_{p+q}(p,\,p-r) \longrightarrow T_{p+q}(p,\,p-1)\}$$
$$= \mathrm{Im}\{T_{p+q}\,F_p/F_{p-r} \longrightarrow T_{p+q}\,F_p/F_{p-1}\}\,.$$

Then we find

$$\cdots \subset Z^3_{pq} \subset Z^2_{pq} \subset Z^1_{pq} = T_{p+q}(p,\,p-1)\,,$$

and if we interpret $F_{-\infty}\,K = 0$ then

$$Z^\infty_{pq} = \mathrm{Im}\{T_{p+q}(F_p) \longrightarrow T_{p+q}\,F_p/F_{p-1}\}\,.$$

Putting

$$B^r_{pq} = \mathrm{Im}\{T_{p+q+1}(p+r-1,\,p) \longrightarrow T_{p+q}(p,\,p-1)\}\,,$$

we have

$$0 = B^1_{pq} \subset B^2_{pq} \subset B^3_{pq} \subset \cdots\,,$$

and if we interpret $F_\infty K = K$, then we have

$$B_{pq}^\infty = \text{Im}\,\{T_{p+q+1}K/F_p K \longrightarrow T_{p+q}F_p/F_{p-1}\}\,.$$

Thus we have

$$0 = B^1 \subset B^2 \subset \cdots \subset B^\infty \subset Z^\infty \subset \cdots \subset Z^2 \subset Z^1 = T_{p+q}F_p/F_{p-1}\,.$$

Now exactness of

$$T_n(p-1, p-r) \longrightarrow T_n(p, p-r) \longrightarrow T_n(p, p-1)$$

implies

$$Z_{pq}^r = \text{Cok}\{T_{p+q}(p-1, p-r) \longrightarrow T_{p+q}(p, p-r)\}$$

and

$$Z_{pq}^{r+1} = \text{Cok}\{T_{p+q}(p-1, p-r-1) \longrightarrow T_{p+q}(p, p-r-1)\}\,.$$

Similarly

$$T_{n+1}(p+r-1, p-1) \longrightarrow T_{n+1}(p+r-1, p) \xrightarrow{\partial} T_n(p, p-1)$$

gives

$$B_{pq}^r = \text{Cok}\{T_{p+q+1}(p+r-1, p-1) \longrightarrow T_{p+q+1}(p+r-1, p)\}$$

and

$$B_{p-r,q+r-1}^r = \text{Cok}\{T_{p+q}(p-1, p-r-1) \longrightarrow T_{p+q}(p-1, p-r)\},$$

and we find

$$
\begin{array}{ccccc}
T_{p+q}(p-1, p-r-1) & \longrightarrow & T_{p+q}(p, p-r-1) & \longrightarrow & Z_{pq}^{r+1} \longrightarrow 0 \\
\downarrow & & \downarrow & & \\
T_{p+q}(p-1, p-r) & \longrightarrow & T_{p+q}(p, p-r) & \longrightarrow & Z_{pq}^{r} \longrightarrow 0 \\
\downarrow & & \downarrow & & \\
B_{p-r,q+r-1}^{r} & & B_{p-r,q+r-1}^{r+1} & & \\
\downarrow & & \downarrow & & \\
0 & & 0 & &
\end{array}
$$

If we set

$$E^r_{pq} = Z^r_{pq} / B^r_{pq}$$

and define

$$d_r : E^r_{pq} \longrightarrow E^r_{p-r,q+r-1}$$

$$\| \qquad\qquad \|$$

$$Z^r_{pq} / B^r_{pq} \longrightarrow Z^r_{pq} / Z^{r+1}_{pq} = B^{r+1}_{pq} / B^r_{pq} \,,$$

then $d_r^2 = 0$ on E^r_{**} and has homology $H(E^r_{**}, d_r) = E^{r+1}_{**}$.

Chapter 23
Spectral Sequences Continued

Recall that

$$T_n(p, p') = T_n(F_p K / F_{p'} K),$$

$$E^1_{p,q} = T_{p+q}(p, p-1).$$

We define

$$0 = B^1_{pq} \subset B^2_{pq} \subset \ldots \subset B^\infty_{pq} \subset Z^\infty_{pq} \subset \ldots \subset Z^2_{pq} \subset Z^1_{pq} = E^1_{pq},$$

$$Z^r_{pq} = \mathrm{Ker}\left\{ T_{p+q}(p, p-1) \xrightarrow{\partial} T_{p+q-1}(p-1, p-r) \right\}$$
$$= \mathrm{Im}\left\{ T_{p+q}(p, p-r) \longrightarrow T_{p+q-1}(p-1, p-r) \right\}$$
$$= \mathrm{Cok}\left\{ T_{p+q}(p-1, p-r) \longrightarrow T_{p+q}(p, p-r) \right\},$$

$$Z^{r+1}_{pq} = \mathrm{Cok}\{ T_{p+q}(p-1, p-r-1) \longrightarrow T_{p+q}(p, p-r-1) \},$$

$$B^r_{pq} = \mathrm{Im}\left\{ T_{p+q+1}(p+r-1, p) \xrightarrow{\partial} T_{p+q}(p, p-1) \right\}$$
$$= \mathrm{Cok}\left\{ T_{p+q+1}(p+r-1, p-1) \longrightarrow T_{p+q+1}(p+r-1, p) \right\},$$

$$B^r_{p-r,q+r-1} = \mathrm{Cok}\left\{ T_{p+q}(p-1, p-r-1) \longrightarrow T_{p+q}(p-1, p-r) \right\},$$

$$B^{r+1}_{p-r,q+r-1} = \mathrm{Cok}\left\{ T_{p+q}(p, p-r-1) \longrightarrow T_{p+q}(p, p-r) \right\}.$$

© Springer Nature Switzerland AG 2020
R. Penner, *Topology and K-Theory*, Lecture Notes in Mathematics 2262,
https://doi.org/10.1007/978-3-030-43996-5_23

We thus have

$$
\begin{array}{ccccccc}
T_{p+q}(p-1, p-r-1) & \longrightarrow & T_{p+q}(p, p-r-1) & \longrightarrow & Z_{pq}^{r+1} & \longrightarrow & 0 \\
\downarrow & & \downarrow & & \cap & & \\
T_{p+q}(p-1, p-r) & \longrightarrow & T_{p+q}(p, p-r) & \longrightarrow & Z_{pq}^{r} & \longrightarrow & 0 \\
\downarrow & & \downarrow & & & & \\
B_{p-r,q+r-1}^{r} & \subset & B_{p-r,q+r-1}^{r+1} & & & & \\
\downarrow & & \downarrow & & & & \\
0 & & 0 & & & &
\end{array}
$$

Thus,

$$
\boxed{Z_{pq}^{r}/Z_{pq}^{r+1} \approx B_{p-r,q+r-1}^{r+1}/B_{p-r,q+r-1}^{r}}
$$

by the previous useful diagram and lemma from last time.

Put $E_{pq}^{r} = Z_{pq}^{r}/B_{pq}^{r}$ and consider

$$
E_{p+r,q-r+1}^{r} \xrightarrow{\ d_r\ } E_{pq}^{r} \xrightarrow{\ d_r\ } E_{p-r,q+r-1}^{r}
$$

$$
\begin{array}{ccccc}
Z_{p+r,q-r+1}^{r} & & Z_{q}^{r} & & Z_{p-r,q+r-1}^{r} \\
\cup & & \cup & & \cup \\
Z_{p+r,q-r+1}^{r+1} & & Z_{pq}^{r+1} & & Z_{p-r,q+r-1}^{r+1} \\
\cup & & \cup & & \cup \\
B_{p+r,q-r+1}^{r+1} & \cong & B_{pq}^{r+1} & \cong & B_{p-r,q+r-1}^{r+1} \\
\cup & & \cup & & \cup \\
B_{p+r,q-r+1}^{r} & & B_{pq}^{r} & & B_{p-r,q+r-1}^{r}
\end{array}
$$

where the boxed result above gives an isomorphism between the quotients of top two terms with the quotients of the adjacent bottom two terms.

Define $d_r : E^r_{p,q} \to E^r_{p-r,q+r-1}$ to be the composition

$$E^r_{pq} \longrightarrow Z^r_{pq}/Z^{r+1}_{pq} \quad \cong \quad B^{r+1}_{p-r,q+r-1}/B^r_{p-r,q+r-1}$$

$$\cap$$

$$E^r_{p-r,q+r-1}$$

Clear that $d^2_r = 0$ and

$$E^{r+1}_{pq} = \frac{\mathrm{Ker}\{d_r : E^r_{pq} \longrightarrow E^r_{p-r,q+r-1}\}}{\mathrm{Im}\{d_r : E^r_{p+r,q-r+1} \longrightarrow E^r_{pq}\}}.$$

We get the spectral sequence

$$E^r_{**} = \{E^r_{pq}\}, \qquad r = 1, 2, \dots ,$$

and on each E^r there is $d_r : E^r_{pq} \to E^r_{p-r,q+r-1}$ so that $d^2_r = 0$, and $E^{r+1} = H(E_r, d_r)$.
   The beginning of the spectral sequence is

$$E^1_{pq} = T_{p+q}(p, p-1)$$
$$= T_{p+q}(F_p K / F_{p-1} K)$$

and the end is

$$E^\infty_{pq} = \frac{\mathrm{Im}\{T_{p+q} F_p K \longrightarrow T_{p+q} K\}}{\mathrm{Im}\{T_{p+q} F_{p-1} K \longrightarrow T_{p+q} K\}}.$$

$T_* K$ is called the *abutment*.

Convergence Question: In what sense is $E^\infty = \lim E^r$?

$$Z^\infty_{pq} = \mathrm{Ker}\left\{T_{p+q}(p, p-1) \xrightarrow{\partial} T_{p+q-1}(F_{p-1} K)\right\} ,$$

and

$$B^\infty_{pq} = \mathrm{Im}\left\{T_{p+q+1}(F_p K) \longrightarrow T_{p+q}(p, p-1)\right\} .$$

We say the spectral sequence *converges strongly* if for a given p and q we have

$$Z^\infty_{pq} = Z^r_{pq} \quad \text{and} \quad B^\infty_{pq} = Z^r_{pq} \quad \text{for } r \text{ large enough} .$$

Exercise Consider a ∂-functor $\{T_n\}$, where the ∂'s lower degree by 1, and a complex with a decreasing filtration $\ldots \supset F_q K \supset F_{q+1} K \supset \ldots$. Show then that we have a spectral sequence E^r_{pq}, for $r \geq 2$, starting with

$$E^2_{pq} = T_{p+q}(F_q K / F_{q+1} K)$$

and with

$$T_{p+q}(F_q/F_{q+1})$$
$$\downarrow$$
$$T_{p+q-1}(F_{q+1}/F_{q+2})$$

abutting to $T_* K$.

Reference: Cartan Seminars.

Examples 2 step filtration: $0 = F_{-1} K \subset F_0 K \subset F_1 K = K$ so

$$E^1_{pq} = T_{p+q} F_p K / F_{p-1} K$$
$$= 0 \quad \text{for } p \neq 0, 1.$$

with

(∗) $\qquad 0 \longrightarrow F_0/F_{-1} \longrightarrow F_1/F_{-1} \longrightarrow F_1/F_0 \longrightarrow 0$

$\qquad\qquad\qquad \| \qquad\qquad\quad \| \qquad\qquad\quad \|$

$\qquad\qquad\qquad K' \qquad\qquad\quad K \qquad\qquad\quad K''$

and we suppose $T_n = 0$ for all $n < 0$.

	$T_3 K'$		
	$T_2 K'$	$T_3 K''$	
q	$T_1 K'$	$T_2 K''$	
	$T_0 K'$	$T_1 K''$	
	$T_{-1} K'$	$T_0 K''$	
		p	

$$d_1 : E^1_{pq} \longrightarrow E^1_{p-1,q}$$
$$\| \qquad\qquad\qquad \|$$
$$T_{p+q}(F_p/F_{p-1}) \xrightarrow{\partial} T_{p+q-1}(F_{p-1}/F_{p-2}),$$

and it can be shown that $d_1 = 0$ for the long exact sequence.
Now d_2 lowers p by 2 and hence is zero, whence $E^\infty_{pq} = E^2_{pq}$.

Thus $T_n K$ has a 2-step filtration

$$
\begin{aligned}
E_{0,n}^{\infty} &= \mathrm{Im}\{T_n K' \to T_n K\} = E_{0,n}^2 \\
&= \mathrm{Cok}\{T_{n+1}(K'') \to T_n K'\} \\
E_{1,n-1}^{\infty} &= T_n K / \mathrm{Im}\{T_n K' \to T_n K\} \\
&= E_{1,n-1}^2 \\
&= \mathrm{Ker}\{T_n K'' \to T_{n-1} K'\}\,.
\end{aligned}
$$

This gives the same information as the long exact sequence of $(*)$.
Another picture of the spectral sequence:

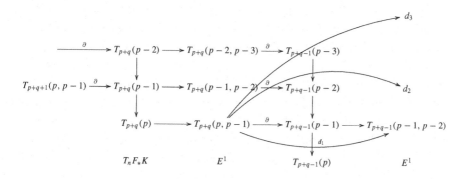

A complex K has two canonical filtrations by dimension:
   (1) the increasing one: $\cdots \to K_{n+1} \to K_n \to \cdots$

$$
(F_p K)_n = \begin{cases} 0, & n > p, \\ K_n, & n \le p, \end{cases}
$$
$$
F_p K / F_{p-1} K = K_p[p]
$$

   (2) the decreasing one (the *Postnikov* filtration):

$$
\begin{array}{ccccccccc}
\cdots & \longrightarrow & K_{n+2} & \longrightarrow & K_{n+1} & \longrightarrow & Z_n & \longrightarrow 0 \longrightarrow 0 \longrightarrow \cdots & F_n K \\
 & & \| & & \| & & \downarrow & & \\
\cdots & \longrightarrow & K_{n+2} & \longrightarrow & K_{n+1} & \longrightarrow & K_n & \longrightarrow K_{n-1} \longrightarrow K_{n-2} \longrightarrow \cdots & K
\end{array}
$$

with

$$
H_i(F_n K) \xrightarrow{\approx} H_i K \quad \text{for} \quad i \ge n
$$

$$
= 0 \quad \text{else}
$$

and

$$F_n K / F_{n+1} K \text{ is given by } 0 \longrightarrow K_{n+1}/Z_{n+1} \longrightarrow Z_n \longrightarrow 0\,,$$

so $F_n K / F_{n+1} K$ is quis to $H_n(K)[n]$.

Chapter 24
Hyper-Homology Spectral Sequences

General Remark 24.1 Let A be a complex. Then we have the exact

$$0 \longrightarrow A \longrightarrow M(A \to 0) \longrightarrow SA \longrightarrow 0$$

$$\{A_n\} \qquad \{A_n \oplus A_{n-1} \oplus 0\} \qquad \{A_{n-1}\}$$

where $M(A \to 0)$ is the mapping cylinder of $A \to 0$ and so is homotopy equivalent to 0, and SA is the suspension. Thus if T is a ∂-functor on $K\,\mathcal{A}$, then we have the exact

$$\cdots \longrightarrow T_n A \longrightarrow T_n M \longrightarrow T_n SA \longrightarrow T_{n-1} A \longrightarrow T_{n-1} M \longrightarrow \cdots$$

$$0 \qquad\qquad\qquad 0$$

so $T_n SA \approx T_{n-1} A$ canonically.

General Remark 24.2 Suppose T is so that $T_n A = 0$ for all $n < 0$. Then

$$T_n A / F_p A = T_{n-(p+1)}(S^{-p-1} A / F_p A)$$
$$= 0 \quad \text{if} \quad n - (p+1) < 0.$$

Thus $T_n A / F_p A = 0$ if $n \le p$, and we have the exact

$$\cdots \longrightarrow T_{n+1} A / F_p A \longrightarrow T_n F_p A \longrightarrow T_n A \longrightarrow T_n A / F_p A \longrightarrow \cdots$$

whence

$$\mathrm{Im}\{T_n F_p A \longrightarrow T_n A\} = T_n A \quad \text{for } p > n \,.$$

© Springer Nature Switzerland AG 2020
R. Penner, *Topology and K-Theory*, Lecture Notes in Mathematics 2262,
https://doi.org/10.1007/978-3-030-43996-5_24

If moreover $F_{<0}A = A$, then

$$T_n A / F_{<0} A = 0 \text{ which implies that } \mathrm{Im}\{T_n A / F_{<0} A \longrightarrow T_n A\} = 0 .$$

Thus the E^1_{pq} spectral sequence is first quadrant and converges.

General Remark 24.3 $L_n F(A) = H_n(F(P))$ is a ∂-functor as in Remark 2, where $P \xrightarrow{\text{quis}} A$ is a projective resolution, since if A is a chain complex, P can also be chosen as a chain complex.

Now, let K be a complex

$$\cdots \longrightarrow K_{n+1} \longrightarrow K_n \longrightarrow \cdots \longrightarrow K_0 \longrightarrow 0$$

and T_n a ∂-functor on $K\,\mathcal{A}$ so that T_* carries quasi-isomorphisms to isomorphisms. We have the two canonical filtrations of last time.

(I) **Increasing:** $F_n K$ is $0 \to \cdots \to 0 \to K_n \to K_{n-1} \to \cdots K_0 \to 0$ and so $F_n K /$
$F_{n-1} K = K_n[n] = S^n K_n[0]$. We get a spectral sequence with

$$E^1_{pq} = T_{p+q}(K_p[p]) = T_q(K_p[0]) \text{ abutting to gr } T_* K .$$

(II) **Decreasing (Postnikov):** $F_n K$ is

$$\mathrm{Ker}\{K_n \xrightarrow{\partial} K_{n-1}\}$$
$$\|$$
$$\cdots \longrightarrow K_{n+2} \longrightarrow K_{n+1} \longrightarrow Z_n \longrightarrow 0 \longrightarrow 0 \longrightarrow \cdots$$

and so $F_n K / F_{n-1} K$ is

$$
\begin{array}{ccccccccc}
0 & \longrightarrow & K_{n+1} & \longrightarrow & Z_{n+1} & \longrightarrow & Z_n & \longrightarrow & 0 \\
 & & \downarrow & & \downarrow & & \downarrow & & \\
0 & \longrightarrow & 0 & \longrightarrow & 0 & \longrightarrow & Z_n/B_n & \longrightarrow & 0 \\
 & & & & & & \| & & \\
 & & & & & & H_n K & &
\end{array}
$$

Thus $F_n K / F_{n+1}$ is quis to $H_n K[n]$. We get a spectral sequence with

$$E^2_{pq} = T_{p+q}(F_q K / F_{q+1} K) = T_p(H_q K[0]) \text{ abutting to gr}(T_* K) .$$

The spectral sequences for $T_{<0} = 0$ of (I) and (II) satisfy Remark 2 hence converge. They are called the *hyper-homology spectral sequences for* $T(K)$.

Program To apply the above to the homology of a small category C, let $\mathrm{Hom}(C, \mathrm{Ab})$ be the category of all functors $F : C \to \mathrm{Ab}$, itself an abelian category. For any $Y \in \mathrm{Ob}\, C$, we have a functor

$$i_Y : \mathrm{point} \longrightarrow C$$
$$: * \longmapsto Y$$

which gives rise to

$$\mathrm{Ab} = \mathrm{Hom}(\mathrm{point}, \mathrm{Ab}) \xleftarrow[\;\;i_{Y!}\;\;]{\;\;i_{Y*}\;\;\;\;i_{Y*}\;\;} \mathrm{Hom}(C, \mathrm{Ab}) \,,$$

and for $A \in \mathrm{Ob}\,\mathrm{Ab}$, we have

$$(i_{Y!}A)(X) = \varinjlim_{(*, i_{Y*} \to X)} A(*) = \bigoplus_{Y \to X} A \,,$$

$$(i_{Y*}(A))(X) = \varprojlim_{(*, i_Y(*) \leftarrow X)} A(*) = \prod_{X \to Y} A$$

by the Kan formulae. Moreover, by the adjunction formulae

(†) $$\mathrm{Hom}_{\mathrm{Hom}(C,\mathrm{Ab})}(i_{Y!}A, F) = \mathrm{Hom}_{\mathrm{Ab}}(A, i_Y^* F)$$
$$= \mathrm{Hom}_{\mathrm{Ab}}(A, F(Y))$$

$$\mathrm{Hom}_{\mathrm{Hom}(C,\mathrm{Ab})}(F, i_{Y*}A) = \mathrm{Hom}_{\mathrm{Ab}}(i_Y^* F, A)$$
$$= \mathrm{Hom}_{\mathrm{Ab}}(F(Y), A) \,.$$

Proposition $\mathrm{Hom}(C, \mathrm{Ab})$ *has enough projectives.*

Proof The functor $F \mapsto i_Y^* F = F(Y)$ is exact. Thus if A is projective, then $F \mapsto \mathrm{Hom}(A, F(Y))$ is exact, so by (†) $i_{Y!}A$ is projective. Dually if A is injective, then $i_{Y*}A$ is as well. Thus, we have the

General Principle If $f : \mathcal{A} \to \mathcal{B}$ is an additive functor between abelian categories having an exact right adjoint g, then f carries projectives into projectives.

Now, note that if we are given a surjection $A \twoheadrightarrow F(Y)$, then the induced map

$$i_{Y!}(A) \longrightarrow F$$

is surjective when applied to Y since

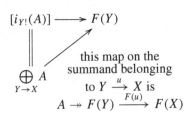

Thus for $X = Y$ with $u = \mathrm{id}_Y$ we get surjectivity.

We conclude that if we choose for each $Y \in \mathrm{Ob}\,\mathcal{C}$ a projective $P_Y \twoheadrightarrow F(Y)$, which is possible since Ab has enough projectives, then we get a surjection

$$\bigoplus_{Y \in \mathrm{Ob}\,\mathcal{C}} i_{Y!}P_Y \twoheadrightarrow F\,,$$

as desired. $\square$

Now, $f : \mathcal{C} \to \mathcal{C}'$ induces

$$f_! : \mathrm{Hom}(\mathcal{C}, \mathrm{Ab}) \longrightarrow \mathrm{Hom}(\mathcal{C}', \mathrm{Ab})\,,$$

which is a right exact functor since $\mathrm{Hom}(f_! F, G) = \mathrm{Hom}(F, f^*G)$ and because of the

Proposition *The sequence $A' \to A \to A'' \to 0$ is exact if and only if for all B,
$0 \to \mathrm{Hom}(A'', B) \to \mathrm{Hom}(A, B) \to \mathrm{Hom}(A', B)$ is, where A, A', A'' and B are
in some abelian category.*

Proof Exercise. $\square$

Thus given $F' \to F \to F'' \to 0$ exact in $\mathrm{Hom}(\mathcal{C}, \mathrm{Ab})$, we get the exact

$$\mathrm{Hom}(F', f^*G) \longleftarrow \mathrm{Hom}(F, f^*G) \longleftarrow \mathrm{Hom}(F'', f^*G) \longleftarrow 0$$
$$\Big\| \qquad\qquad\qquad \Big\| \qquad\qquad\qquad \Big\|$$
$$\mathrm{Hom}(f_! F', G) \longleftarrow \mathrm{Hom}(f_! F, G) \longleftarrow \mathrm{Hom}(f_! F'', G) \longleftarrow 0$$

so that $f_! F' \to f_! F \to f_! F''$ is exact.

In particular, if $f : \mathcal{C} \to \mathrm{point}$, then

$$(f_! F)(*) = \varinjlim_{(X, fX \to *)} F(X) = \varinjlim_{\mathcal{C}} A(X)\,.$$

Now given any complex $F.$ bounded below, we choose a projective resolution $P. \to F.$ and define

$$(L_q f_!)(F.) = H_q\{f_! P.\}$$

where $f : \mathcal{C} \to \mathcal{C}'$, $P.$ and $F.$ in $C_+(\mathrm{Hom}(\mathcal{C}, \mathrm{Ab}))$, $f_! (P.)$ in $C_+(\mathrm{Hom}(\mathcal{C}', \mathrm{Ab}))$ and $(L_q f_!)(F.)$ in $\mathrm{Hom}(\mathcal{C}', \mathrm{Ab})$ for each q. We regard F in $\mathrm{Hom}(\mathcal{C}, \mathrm{Ab})$ as the complex $F[0]$, and we write

$$(L_q f_!)(F) = (L_q f_!)(F[0]) .$$

Finally, $T_n = L_n f_!$ is a ∂-functor which inverts quasi-isomorphisms, so we get the hyper-homology spectral sequences convergent by Remark 3 with

$$E^1_{pq} = (L_q f_!)(F_p) \text{ converging to } \mathrm{gr}\, L_* f_!(F.)$$

and

$$E^2_{pq} = (L_p f_!)(H_q(F.)) \text{ converging to } \mathrm{gr}\, L_* f_!(F.) .$$

Example (*where these sequences collapse*) Assume that F_q is acyclic for $f_!$, i.e.,
$$(L_q f_!)(F_p = \begin{cases} 0, & \text{for } q \neq 0, \\ f_! F_q, & \text{for } q = 0. \end{cases}$$
We find

and check easily that

$$d^1 : E^1_{pq} \longrightarrow E^1_{p-1,q}$$
$$\parallel \qquad\qquad \parallel$$
$$T_q F_p \qquad\quad T_q F_{p-1}$$

is induced by $F_p \xrightarrow{d} F_{p-1}$.

Thus d^1 is the only non-zero differential, and

$$E^1_{*0} = f_! F_* ,$$
$$E^1_{*q} = 0 \quad \text{for } q \neq 0,$$

so

$$E_{pq}^2 = \begin{cases} 0, & \text{if } q \neq 0, \\ H_p(f_! F.), & \text{if } q = 0, \end{cases}$$

and we conclude

$$(L_p f_!)(F.) = H_p(f_! (F.))$$

whenever all F_n are acyclic for $f_!$.

Chapter 25
Generalized Kan Formulae

Suppose that $C \xrightarrow{f} C'$ is a functor between small categories. We have

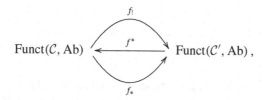

$$\text{Funct}(C, \text{Ab}) \qquad \qquad \text{Funct}(C', \text{Ab}),$$

and by the Kan formula

$$(f_! F)(Y) = \varinjlim_{(X, fX \to Y) \in f/Y} F(X).$$

Moreover, the adjunction formula says that

$$(*) \qquad \qquad \text{Hom}(f_! F, G) = \text{Hom}(F, f^* G),$$

and we saw that F projective implies $f_! F$ projective.

Let $\mathcal{A} = \text{Funct}(C, \text{Ab})$, $M. \in C_+(\mathcal{A})$, and $\mathbb{L} f_!(M.) = f_!(P.)$, where $P. \to M.$ is a projective resolution, and then $(L_q f_!)(M.) = H_q(\mathbb{L} f_!(M.))$ is a ∂-functor carrying quasi-isomorphisms to isomorphisms. This last is true since if we have

$$
\begin{array}{ccc}
P. & \xrightarrow{\text{quis}} & M. \\
& & \downarrow{\text{quis}} \\
P'. & \xrightarrow{\text{quis}} & M'.
\end{array}
$$

R. Penner, *Topology and K-Theory*, Lecture Notes in Mathematics 2262,
https://doi.org/10.1007/978-3-030-43996-5_25

then $P_{\cdot}$ is a projective resolution of M', and the map $P_{\cdot} \to P'$ making the diagram commute is a homotopy equivalence.

Recall the two spectral sequences

$$E^1_{pq} = L_q f_!(M_p) \text{ converging to } \mathrm{gr}\{L_* f_!(M_{\cdot})\},$$
$$E^2_{pq} = L_p f_!(H_q(M_{\cdot})) \text{ converging to } \mathrm{gr}\{L_* f_!(M_{\cdot})\}.$$

Note that if $F \in \mathcal{A}$, then

$$L_q f_!(F) = L_q f_!(F[0])$$
$$= \begin{cases} f_! F, & \text{if } q = 0, \\ 0, & \text{if } q < 0. \end{cases}$$

This is true since we construct

$$\cdots \longrightarrow P_1 \longrightarrow P_0 \longrightarrow F \longrightarrow 0$$

and

$$L_q f_!(F) = H_q(\cdots \longrightarrow f_! P_1 \longrightarrow f_! P_0 \longrightarrow 0 \longrightarrow \cdots)$$
$$= f_! F$$

in degree zero since $f_!$ is right exact by the adjunction formula (*).

If all the M_p are acyclic for $f_!$, then the first spectral sequence collapses and gives

$$E^1_{pq} = \begin{cases} 0, & \text{if } q \neq 0, \\ f_! M_p, & \text{if } q = 0, \end{cases}$$

and

$$E^\infty_{pq} = E^2_{pq} = \begin{cases} 0, & \text{if } q \neq 0, \\ H_p(f_! M_{\cdot}), & \text{if } q = 0, \end{cases}$$

so that $H_p(f_! M_{\cdot}) = L_p f_!(M_{\cdot})$.

In other words, when all the M_p are $f_!$-acyclic, we have that the projective resolution $P_{\cdot} \to M_{\cdot}$ induces a quis $f_! P_{\cdot} \to f_! M_{\cdot}$, and so in this case we find the quis $\mathbb{L} f_!(M_{\cdot}) \to f_!(M_{\cdot})$.

Consider now $C \xrightarrow{f} C' \xrightarrow{g} C''$. We claim that $g_! \, f_! = (gf)_!$. This follows since

$$\mathrm{Hom}((gf)_! \, F, A) = \mathrm{Hom}(F, (gf)^* A)$$
$$= \mathrm{Hom}(F, f^* g^* A)$$
$$= \mathrm{Hom}(g_! \, f_! \, F, A),$$

and so Yoneda implies that $(gf)_! (F) \cong g_! f_! (F)$.

Suppose we have $C \xrightarrow{f} C' \xrightarrow{g}$ point. Then

$$g_! G = \varinjlim_{C'} G,$$

$$(gf)_! F = \varinjlim_{C} F.$$

In this case we write $H_q(C, F)$ instead of $L_q(gf)_! (F)$ and $H_q(C', G)$ instead of $L_q g_! (G)$.

Suppose $M. \in C_+(\mathrm{Func}(C, \mathrm{Ab}))$ and choose $P. \xrightarrow{\mathrm{quis}} M.$. Then

$$\mathbb{L}(gf)_! (M.) = (gf)_! (P.)$$
$$= g_! f_! (P.).$$

Now, $f_!(P.)$ is a complex of projectives, and projectives are acyclic for $f_!$. Thus

$$\mathbb{L} g_! (f_! P.) \xrightarrow{\mathrm{quis}} g_! (f_! P.).$$

We conclude that

$$\boxed{\mathbb{L}(gf)_! (M.) \xleftarrow{\mathrm{quis}} \mathbb{L} g_! (\mathbb{L} f_! (M.))}$$

Taking homology, we get

$$H_q(\mathbb{L}(gf)_! (M.)) = L_q(gf)_! (M.) = H_q(C, M.),$$

so we find

$$H_q(C, M.) = H_q(C', \mathbb{L} f_! (M.)).$$

Thus the spectral sequence resulting from the Postnikov filtration is

$$E^2_{pq} = H_p(C', L_q f_! (M.)) \text{ converging to } \mathrm{gr}\{H_*(C', \mathbb{L} f_! (M.))\},$$

and we obtain the *Leray spectral sequence in homology* for $f : C \to C'$ with

$$E^2_{pq} = H_p(C', L_q f_! (M.)) \text{ converging to } \mathrm{gr}\, H_*(C, M.).$$

(†) **Problem** Generalize the Kan formula to $(L_q f_! (M.))(Y) = H_q(f/Y, i^*M.)$, where $\mathrm{Ob}\,(f/Y) = \{(X, fX \xrightarrow{u} Y)\}$ and $i : f/Y \to C$ sends (X, u) to X.

Now,

$$
\begin{aligned}
L_q \, f_! \, (M.)(Y) &= H_q(f_! \, P.)(Y) \\
&= H_q((f_! \, P.)(Y)) \\
&= H_q(\varinjlim_{(X,u)\in f/Y} P.(X)) \\
&= H_q(\varinjlim_{f/Y} i^* P.) .
\end{aligned}
$$

Moreover,

$$
H_q(f/Y, i^* M.) = H_q(\varinjlim_{f/Y} Q.) ,
$$

where Q is a projective resolution of $i^* M.$ in $C_+(\mathrm{Func}(f/Y, \mathrm{Ab}))$, thus the problem (†) becomes:

Problem If $P. \to M.$ is a projective resolution, then $i^* P. \to i^* M.$ is a quis since i^* is an exact functor. Is $i^* P.$ a complex of projectives?

If so, then take $Q. = i^* P.$. Note that it would be enough to know that each $i^* P_q$ is acyclic for $\varinjlim_{f/Y}$.

Claim That i^* carries projectives to projectives. This would follow if we could exhibit an exact right adjoint to i^* as before. We show that i_* is exact. Recall that

$$
\begin{aligned}
i : f/Y &\longrightarrow \mathcal{C} \\
: (X, u) &\longmapsto X .
\end{aligned}
$$

Clearly then f/Y is the fibered category with discrete fibers over $\mathcal{C}$ associated to the contravariant functor $X \to \mathrm{Hom}_{\mathcal{C}}(fX, Y)$. Recall that if i is fibered, then

$$
\begin{aligned}
(i_* F)(X) &= \varprojlim_{(Z,X\xrightarrow{u} iZ)\in X\backslash i} F(Z) \\
&= \varprojlim_{Z\in i^{-1}X} F(Z) \\
&= \prod_{Z\in \mathrm{Ob}\, i^{-1}(X)} F(Z) \quad \text{since the fibers are discrete,}
\end{aligned}
$$

which is clearly exact in F. This proves the claim and the

General Fact If $\widetilde{\mathcal{C}} \xrightarrow{i} \mathcal{C}$ is a fibered category with discrete fibers, then i_* is exact and i^* carries projectives to projectives.

We have in fact proved the

Generalized Kan Formula $L_q \, f_! \, (M.)(Y) = H_q(f/Y, i^* M.)$.

Now recall that if f is pre-cofibered, then the Kan formula simplifies to

$$(f_! F)(Y) = \varinjlim_{(X, fX \to Y)} F(X)$$
$$= \varinjlim_{X \in f^{-1}Y} F(X).$$

Immediate Goal If f is pre-cofibered, then

$$L_q f_!(M.)(Y) = H_q(f/Y, i^*M.)$$
$$= H_q(f^{-1}Y, M. \text{ restricted to } f^{-1}(Y))$$
$$= H_q(f^{-1}Y, j^*(i^*M.)).$$

Recall that we have

$$u_* X \longleftarrow \!\!\!\mid (f\overset{X}{X} \overset{u}{\to} Y)$$

$$f^{-1}Y \quad \overset{r}{\underset{j}{\rightleftarrows}} \quad f/Y$$

$$u \longmapsto (u, fu \overset{\text{id}}{=} Y).$$

Thus when f is pre-cofibered, the embedding $j : f^{-1}Y \to f/Y$ has left adjoint r.

For the immediate goal, it suffices to prove that

$$H_q(f/Y, G.) = H_q(f^{-1}Y, j^*G.).$$

$$\boxed{\textbf{Suppose that } j^* = r_!}$$

Then

$$H_p(f/Y, G.) = H_p(f^{-1}Y, \mathbb{L}\, r_! \, G.)$$
$$(\text{since } r_! = j^* \text{ is exact}) = H_p(f^{-1}Y, r_! \, G.)$$
$$(\text{by supposition}) = H_p(f^{-1}Y, j^*G.),$$

as desired.

Whether the boxed supposition is correct will be determined next time.

Chapter 26
The Hochschild–Serre Spectral Sequence

Suppose we have adjoint functors $\mathcal{C} \underset{f}{\overset{g}{\rightleftarrows}} \mathcal{C}'$. Recall that

$$h^X : T \mapsto \operatorname{Hom}_{\mathcal{C}}(X, T)$$

and consider the category $\widehat{\mathcal{C}} = \operatorname{Func}(\mathcal{C}, \operatorname{Sets})$.

Claim $g_! \, h^X = h^{g^X}$.

Proof

$$\operatorname{Hom}_{\widehat{\mathcal{C}'}}(g_! \, h^X, G) = \operatorname{Hom}_{\widehat{\mathcal{C}}}(h^X, g^*G)$$
$$\text{(by Yoneda)} = (g^*G)(X)$$
$$= G(gX)$$
$$\text{(by Yoneda)} = \operatorname{Hom}_{\widehat{\mathcal{C}'}}(h^{g^X}, G)$$

and finally, by Yoneda again $g_! \, h^X = h^{g^X}$. □

Now,

$$(g_! \, h^X)(Y) = h^{g^X}(Y) = \operatorname{Hom}_{\mathcal{C}}(gX, Y)$$
$$= \operatorname{Hom}_{\mathcal{C}}(X, fY)$$
$$= h^X(fY)$$
$$= (f^*h^X)(Y),$$

and therefore $g_! \, h^X = f^*h^X$, so $g_!$ and f^* agree on functors of the form h^X.

© Springer Nature Switzerland AG 2020
R. Penner, *Topology and K-Theory*, Lecture Notes in Mathematics 2262,
https://doi.org/10.1007/978-3-030-43996-5_26

Now, let $F \in \widehat{\mathcal{C}}$ be any functor. Then form the category

$$\mathcal{C}_F = \{(X, \xi) : \xi \in F(X)\},$$

i.e., $\mathcal{C}_F$ is the cofibered category with discrete fibers belonging to $F.$. Then Yoneda says $\xi \in F(X) = \mathrm{Hom}(h^X, F)$, and we claim

$$\varinjlim_{(X,\xi)\in\mathcal{C}_F} h^X \xrightarrow{\sim} F.$$

The proof of this is an exercise as follows. Evaluate on an object Y and identify, or use Yoneda and look at $\mathrm{Hom}(F, G)$.

Now, $g_!$ preserves inductive limits because it has a right adjoint since

$$\mathrm{Hom}(g_! \varinjlim_i F_i, G) = \mathrm{Hom}(\varinjlim_i F_i, g^*G)$$
$$= \varprojlim \mathrm{Hom}(F_i, g^*G)$$
$$= \varprojlim \mathrm{Hom}(g_! F_i, G)$$
$$= \mathrm{Hom}(\varinjlim g_! F_i, G),$$

and finally, by Yoneda, the claim follows.

By the same argument, f^* preserves inductive limits.

Putting all this together,

$$g_!(F) = f^*(F), \quad \text{for all } F \in \widehat{\mathcal{C}}.$$

Thus, given $\mathcal{C} \overset{g}{\underset{f}{\rightleftarrows}} \mathcal{C}'$, we have

In this situation where $g_! = f^*$, then $g_!$ on abelian group-valued functors is the same as $g_!$ on set valued functors. If F is abelian group-valued, then we have

$$+ : F \times F \longrightarrow F$$

and

$$g_!(F \times F) \longrightarrow g_! F$$

and this map is an isomorphism when $g_! = f^*$ since f^* preserves products.

$$g_! F \times g_! F$$

For derived functors, $\mathbb{L} g_!$ is quis to $g_!$ since on $\mathrm{Func}(\mathcal{C}, \mathrm{Ab})$, $g_!$ is exact.

Suppose $\mathcal{C} \xrightarrow{f} \mathcal{C}'$ and $M. \in C_+(\mathrm{Func}(\mathcal{C}, \mathrm{Ab}))$. Last time we saw that

$$(*) \qquad H_n(\mathcal{C}', \mathbb{L} f_!(M.)) = H_n(\mathcal{C}, M.).$$

Moreover, we can show that

$$\mathbb{L} \varinjlim_{\mathcal{C}'} \circ \mathbb{L} f_!(M.) = \mathbb{L} \varinjlim_{\mathcal{C}} (M.).$$

On the other hand,

$$E^2_{pq} = H_p(\mathcal{C}', L_q f_!(M.)) \text{ converges to } \mathrm{gr}\{H_n(\mathcal{C}', \mathbb{L} f_!(M.))\},$$

and we proved the generalized Kan formula

$$L_q f_!(M.)(Y) = H_q(f/Y, i^* M.)$$

where $i : f/Y \to \mathcal{C}$ is the natural map.

Now suppose f is pre-cofibered. We have

$$(\dagger) \qquad f^{-1}(Y) \underset{j}{\overset{r}{\rightleftarrows}} f/Y$$

where j is the embedding, $r(X, fX \xrightarrow{u} Y) = u^* X$ and r is left adjoint to j so that $r_! = j^*$ and $\mathbb{L} r_! = r_!$, thus resolving a question from last time.

Thus, by $(*)$ applied to $(\dagger)$, we get

$$H_n(f/Y, i^* M.) = H_n(f^{-1}(Y), \mathbb{L} r_!(i^* M.))$$
$$= H_n(f^{-1}(Y), j^* i^* M.).$$

Notice that $j^* i^* M.$ is just $M.$ restricted to $f^{-1}(Y)$.

Conclusion When f is pre-cofibered, we have

$$L_q f_!(M.)(Y) = H_q(f^{-1}(Y), M. \text{ restricted to } f^{-1}(Y)).$$

Exercise

$$\mathbb{L} \, f_!(M.)(Y) = \mathbb{L} \varinjlim_{f^{-1}(Y)} (M. \text{ restricted to } f^{-1}(Y)) \, .$$

Example Given a group extension

$$* \longrightarrow N \longrightarrow G \longrightarrow Q \longrightarrow * \, ,$$

we have $\widetilde{G} \xrightarrow{f} \widetilde{Q}$, and f is clearly cofibered and fibered with fiber $f^{-1}(*) = \widetilde{N}$, where $\widetilde{X}$ is the category associated to each $X = G, Q, N$. As usual, a G-module is a functor from $\widetilde{G}$ to Ab. We have

$$\begin{aligned} H_*(\widetilde{G}, M) &= \text{group homology of } G \text{ with values in } M \\ &= H_*(G, M) \, , \end{aligned}$$

and we get a spectral sequence

$$E_{pq}^2 = H_p(\widetilde{Q}, L_q \, f_!(M)) \quad \text{converging to} \quad \text{gr } H_n(\widetilde{G}, M.) = \text{gr } H_n(G, M)$$

and

$$H_p(\widetilde{Q}, L_q \, f_!(M)) = H_p(Q, H_q(N, M)) \, .$$

This is the *Hochschild–Serre spectral sequence*

$$E_{pq}^2 = H_p(Q, H_q(N, M)) \text{ converging to gr } H_n(G, M) \, .$$

Example Take a map $K \to L$ of simplicial complexes and consider the functor

$$\text{Simp}(K) \xrightarrow{f} \text{Simp}(L) \, ,$$

where we take geometric realization of the poset $\text{Simp}(K)$ of simplices in K under inclusion to get the barycentric subdivision, as before.
Then

$$H_*(\text{Simp}(K), A) = H_*(K, A) \, ,$$

where the left-hand side has an abelian group A regarded as a constant functor, and the right-hand side is the usual homology of the simplicial complex with coefficients in A. Associated to f is a spectral sequence

$$E_{pq}^2 = H_p(\text{Simp}(L), L_q \, f_!(A \text{ on Simp } K))$$

converging to $H_n(\text{Simp } K, A \text{ on Simp}(K))$. We showed before that f is fibered, and the fiber is

$$f^{-1}(T) = \{\text{all simplexes mapping onto } T\}.$$

(Open) **Problem** Show that the simplicial complex belonging to $f^{-1}(T)$ is a triangulation of the inverse image of any interior point of T.

If so, then we have the spectral sequence

$$E^2_{pq} = H_p(\text{Simp } L, T \mapsto H_q(f^{-1}(T), A)) \text{ converging to } H_p(\text{Simp } K, A).$$

Here we use contravariant functors, whence we have

$$L_q \, f_!(F)(Y) = H_q(Y \backslash f, F)$$
$$(\text{for } f \text{ pre-fibered}) = H_q(f^{-1}(Y), F).$$

Variants

(1) contravariant functors,

(2) cohomology: for $f : \mathcal{C} \to \mathcal{C}'$, we have $f_* : \text{Func}(\mathcal{C}, \text{Ab}) \to \text{Func}(\mathcal{C}', \text{Ab})$ which is left exact, i.e., $0 \to F' \to F \to F''$ exact implies $0 \to f_*F' \to f_*F \to f_*F''$ exact. Using **injective** resolutions, define

$$\mathbb{R} \, f_*(M.), \quad \text{for } M \in C^*(\text{Func}(\mathcal{C}, \text{Ab})).$$

Exercise Figure out when it happens that

$$\mathbb{R} \, f_*(M.)(Y) = H^q(f^{-1}(Y), M).$$

Let us prove (c). The multiplicative group κ^* of the finite field κ is cyclic of order m prime to p. Hensel lemma shows that the equation $x^m = 1$ has m distinct roots in U_K, from which we deduce that the following exact sequence of abelian groups:

$$1 \longrightarrow U_K^1 \longrightarrow U_K \longrightarrow \kappa^* \longrightarrow 1$$

splits. Hence $U_K \simeq U_K^1 \times \kappa^*$, and by (b) we obtain $K^* \simeq \mathbf{Z} \times \mathbf{Z}_p^n \times F \times \kappa^*$.

We have the following complement to part (b) of Theorem 7.18:

Corollary 7.19 *Let K be a p-adic field. Then for all $n > 0$, the subgroup K^{*n} of nth powers is open in K^*. Every finite index subgroup of K^* is open.*

Proof The second assertion follows from the first, since an index n subgroup of K^* contains K^{*n}. Let us prove the first assertion. For n prime to p, it follows immediately from Hensel lemma. In the case where n is a power of p, the result follows from Theorem 7.18, (b). □

Remark 7.20 If K is a local field of characteristic p, the subgroup K^{*p} is no longer open (it is still closed by compactness of U_K, since $K^* \simeq \mathbf{Z} \times U_K$), otherwise it would contain U_K^m for m large enough. This is clearly not the case since the equation $1 + \pi^m = x^p$ implies $\pi^m = (x - 1)^p$, which is not possible if the valuation m of the left hand term is not divisible by p. On the other hand, one can show that the group U_K^1 is isomorphic to $\mathbf{Z}_p^N$ (Exercise 7.4). In fact in this case K^* contains finite index subgroups which are not closed, cf. Exercise 11.3.

Nevertheless, Hensel lemma still implies that for n prime to p, the subgroup K^{*n} is open in K^*.

7.6 Exercises

Exercise 7.1 Let K be a local field with ring of integers $\mathcal{O}_K$. Let $\mathfrak{m}_K$ be the maximal ideal of $\mathcal{O}_K$.

(a) Show that an element x of K is in $\mathfrak{m}_K$ if and only if for all $n > 0$ not divisible by p, there exists a $y \in K^*$ such that $1 + x = y^n$.

(b) Deduce that any field morphism f from K to K is continuous (show first that for all $s > 0$, the ideal $\mathfrak{m}_K^s$ of $\mathcal{O}_K$ is stable by f).

(c) What are the automorphisms of $\mathbf{Q}_p$?

Exercise 7.2 Let $d \in \mathbf{Z}$ be a squarefree integer. Let p be an odd prime number.

(a) Show that $\mathbf{Q}_p(\sqrt{d})$ is an unramified extension of $\mathbf{Q}_p$ if and only if p does not divide d.

(b) Show that d is a square in $\mathbf{Q}_2$ if and only if d is congruent to 1 modulo 8 (for the "if" part, you may use a formal series F with coefficients in $\mathbf{Q}$ such that $F^2 = 1 + X$).

(c) Show that $\mathbf{Q}_2(\sqrt{d})$ is an unramified extension of $\mathbf{Q}_2$ if and only if d is congruent to 1 modulo 4.

(d) Let $p_1, \ldots, p_r$ be pairwise distinct prime numbers congruent to 1 modulo 4. Let ℓ be any prime number. Show that the ramification index of the extension $\mathbf{Q}_\ell(\sqrt{-p_1 \cdots p_r}, \sqrt{p_1})/\mathbf{Q}_\ell$ is at most 2.

Exercise 7.3 Let K be a local field with residue field κ. Let $\bar{\kappa}$ be the algebraic closure of the finite field κ.

(a) Assume that $K = \kappa((t))$ is a function field. Choose an infinite set $\{a_0, \ldots, a_n, \ldots\}$ of elements of $\bar{\kappa}$. Show that the sequence (u_n) defined by $u_n = \sum_{k=0}^{n} a_k t^k$ is a sequence of elements of K_{nr} which does not converge in K_{nr}. Deduce from it that K_{nr} is not complete (for the discrete valuation that extends that of K).

(b) What is the completion of K_{nr} for $K = \kappa((t))$?

(c) Assume now that K is a p-adic field. Show that K_{nr} is not complete.

(d) Let K be a local field with separable closure $\bar{K}$ and valuation v. We endow $\bar{K}$ with the (non discrete) valuation $v : \bar{K} \to \mathbf{Q} \cup \{\infty\}$ obtained by passing to the limit over finite separable extension L of K (each L endowed with the valuation $v_L : L \to (\frac{1}{e}\mathbf{Z}/\mathbf{Z}) \cup \{\infty\}$ which extends v, where e is the ramification index of L over K). Show that $\bar{K}$ is not complete for this valuation (observe that any sequence of elements of K_{nr} that converges in $\bar{K}$ has its limit in a finite extension L of K, and then show that this limit must remain in K_{nr}).

Exercise 7.4 (*after* [55], *Chap. II.3, Prop. 10*) Let $K = \kappa((t))$ be a local field of characteristic $p > 0$, with valuation v. Fix a uniformiser π of K and let $\mathfrak{m}$ be its maximal ideal. Let $(\alpha_1, \ldots, \alpha_f)$ be a basis of κ over $\mathbf{F}_p$. The aim of this exercise is to determine the structure of the multiplicative group $U_K^1 = \{x \in K, v(1 - x) > 0\}$.

(a) Show that U_K^1 can be endowed with a structure of a $\mathbf{Z}_p$-module by the formula

$$a \cdot x := x^{\sum_{n \geqslant 0} a_n p^n} = \prod_{n \geqslant 0} x^{a_n p^n}$$

for any $x \in U_K$ and any $a = \sum_{n \geqslant 0} a_n p^n$ (with $a_n \in \mathbf{Z}$) in $\mathbf{Z}_p$.

(b) Let $N > 0$, set $N = np^\nu$ with n not divisible by p and $\nu \geqslant 0$. Let $a_1, \ldots, a_f \in \mathbf{N}$. Prove the formula

$$\prod_{i=1}^{f}(1 + \alpha_i \pi^n)^{a_i p^\nu} \equiv 1 + \left(\sum_{i=1}^{f} a_i \beta_i\right)\pi^N \bmod.\mathfrak{m}^{N+1},$$

where $\beta_i := \alpha_i^{p^\nu}$.

(c) Let x_1 be any element of $\mathfrak{m}$. Show that one can inductively define a sequence (x_N) with $x_N \in \mathfrak{m}^N$ for any $N > 0$, by setting $1 + x_{N+1} = (1 + x_N)y_N^{-1}$, where y_N is defined by the formula

Chapter 27
Resolution for Exact Categories

Consider an additive full subcategory $\mathcal{M}$ of an additive category $\mathcal{A}$ with its induced notion of exact sequences $0 \to M' \to M \to M'' \to 0$ inherited from $\mathcal{A}$. Then $\mathcal{M}$ is an *exact category* provided it is closed under extensions in $\mathcal{A}$, i.e., if M', M'' are objects in $\mathcal{M}$, then so too is M. We can cook up an axiomatic notion of such, but it is not worth it here. Note that the set of projective modules is an exact but not an abelian category.

Example (of properties of exact sequences)

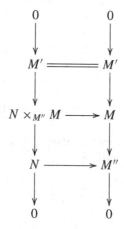

Then the vertical left-hand sequence is exact if the right-hand one is because $\mathcal{M}$ is closed under extensions $N \times_{M''} M \in \mathcal{M}$ and similarly for push-outs of the first term of the exact sequence.

© Springer Nature Switzerland AG 2020
R. Penner, *Topology and K-Theory*, Lecture Notes in Mathematics 2262,
https://doi.org/10.1007/978-3-030-43996-5_27

Given an exact category $\mathcal{M}$, a map $M \to M'$ is an *admissible monomorphism* if it is part of an exact sequence $0 \to M \to M' \to M'' \to 0$ in $\mathcal{M}$, and similarly for an *admissible epimorphism*. An *admissible filtration* of M is a filtration

$$0 = F_{-1}M \subset F_0 M \subset \ldots \subset F_n M = M$$

so that each $F_{p-1}M \subset F_p M$ is an admissible monomorphism.

Note that composition of admissible monomorphisms is admissible since the cokernel of the composition is an extension of the respective cokernels.

Resolution Theorem *Given an exact category $\mathcal{M}$ and a full subcategory $\mathcal{P}$ closed under extensions in $\mathcal{M}$. Assume that*
(i) if $0 \to M' \to P \to P' \to 0$ is exact in $\mathcal{M}$ and $P, P' \in \mathcal{P}$ implies also that $M' \in \mathcal{P}$; note that the kernel of maps between projectives is itself projective.
(ii) for any $M \in \mathcal{M}$, there exists an admissible exact sequence

$(*)$ $\qquad\qquad 0 \longrightarrow P_n \longrightarrow \cdots \longrightarrow P_1 \longrightarrow P_0 \longrightarrow M \longrightarrow 0$

so that $P_i \in \mathcal{P}$, where admissible *here means that each associated short exact sequence is admissible.*
    Then $K_0 \mathcal{P} \to K_0 \mathcal{M}$ is an isomorphism.

Proof It remains (from last time) to show that given $(*)$, then

$$\sum (-1)^i [P_i] \in K_0 \mathcal{P}$$

is independent of the resolution.

Suppose we have resolutions

$$0 \longrightarrow P_n \longrightarrow \cdots \longrightarrow P_0 \longrightarrow M \longrightarrow 0$$

$$0 \longrightarrow P'_n \longrightarrow \cdots \longrightarrow P'_0 \longrightarrow M \longrightarrow 0$$

where without loss of generality we take them to have same length by adding zeros if necessary. Construct the fiber product

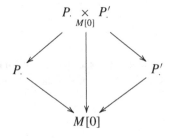

and get another resolution of M, which may however not be in $\mathcal{P}$; if it were, then we would be done by considering the kernel of $P. \underset{M[0]}{\times} P' \to P.$ and showing it is in $\mathcal{P}$ using (i) above.

More formally, we require the

Lemma *Consider an exact sequence*

$$0 \longrightarrow K. \longrightarrow Q. \longrightarrow P. \longrightarrow 0,$$

of finite complexes in $\mathcal{M}$*, where* $\mathcal{M}$ *satisfies condition (i) above, so that each* Q_i *and* P_i *are in* $\mathcal{P}$ *and so that* $K.$ *is acyclic. Then we have*

$$\sum (-1)^i [Q_i] = \sum (-1)^i [P_i] \in K_0 \mathcal{P}.$$

Proof Condition (i) implies that each $K_i \in \mathcal{P}$ since $0 \to K_n \to \cdots \to K_0 \to 0$ is exact, and $K.$ is admissible in $\mathcal{P}$, so

$$\sum (-1)^i [K_i] = 0 = \sum (-1)^i ([Q_i] - [P_i]) . \qquad \square$$

Now, define $\mathcal{P}_n$ to be the subcategory of $\mathcal{M}$ consisting of elements of $\mathcal{M}$ having a resolution (*) of length at most n with each $P_i \in \mathcal{P}$. We therefore have $\mathcal{P} = \mathcal{P}_0 \subset \mathcal{P}_1 \subset \cdots \subset \bigcup_n \mathcal{P}_n = \mathcal{M}$.

We prove $K_0 \mathcal{P}_0 \xrightarrow{\sim} K_0 \mathcal{P}_1$, i.e., suppose $M \in \mathcal{P}_1$ and we have

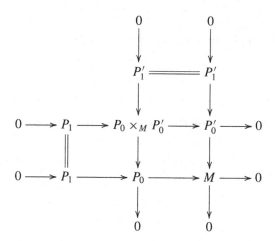

Thus $P_0 \times_M P_0'$ is an extension of P_1 by P_0', and hence $P_0 \times_M P_0' \in \mathcal{P}$, as desired. Clearly

$$[P_0] - [P_1] = [P_0] - [P_0 \times_M P_0'] + [P_0']$$
$$= [P_0'] - [P_1'].$$

To finish the argument, we must show that $\mathcal{P}_n$ is closed under extensions, that $0 \to M' \to M \to M'' \to 0$ is exact in $\mathcal{P}_n$ and also that $M'', M \in \mathcal{P}_{n-1}$ and $M' \in \mathcal{P}_n$ implies $M' \in \mathcal{P}_{n-1}$. See Quillen's paper; these are straightforward but tedious.

This proves the theorem. $\square$

Here is the proof that $\mathcal{P}_n$ is closed under extensions for $n = 2$. Suppose we have the diagram

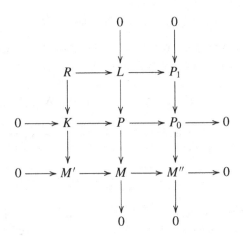

where $K, P \in \mathcal{P}$ with $0 \to M' \to M \to M'' \to 0$ exact. It suffices to show that $R \in \mathcal{P}$, i.e.,

$$0 \longrightarrow R \longrightarrow P \longrightarrow M' \longrightarrow 0$$

with $P \in \mathcal{P}$ and $M' \in \mathcal{P}_1$ implies that $R \in \mathcal{P}$. But we have

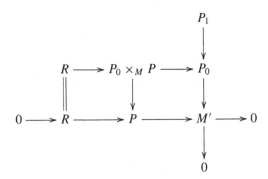

so indeed $R \in \mathcal{P}$ as desired.

Corollary *Let A be a regular noetherian ring, i.e., every finitely generated module has a finite resolution by finitely generated projective A-modules, e.g., $A = k[T_1, \dots, T_n]$ by the Szyzygy Theorem.*

Then recall that $K_0 A = K_0 \mathcal{P}_A$. *By the Resolution Theorem,* $K_0 \mathcal{P}_A \xrightarrow{\sim} K_0$ Modf(A). *(This is also true for* $A[T]$; *the proof of Szyzygy Theorem says* $A[T]$ *is regular, Quillen thinks.)*

Example If A is noetherian, then the collection of finitely generated A-modules is an abelian category. Let S be a multiplicative system in A and form $S^{-1}A$. We have

$$
\begin{array}{ccccc}
 & & \mathcal{A} & & \\
 & & \| & & \\
\{S\text{-torsion finitely generated } A\text{-modules}\} & \longrightarrow & \text{Modf}(A) & \longrightarrow & \text{Modf } S^{-1}A \\
\| & & & & \\
\mathcal{B} & & & M & \longmapsto & S^{-1}M
\end{array}
$$

A *thick* (or *Serre*) subcategory of an abelian category $\mathcal{A}$ is a full subcategory which is closed under sub-objects, quotient objects and extensions, for example, $\mathcal{B} \subset \mathcal{A}$.

Now, given a thick subcategory $\mathcal{B} \subset \mathcal{A}$, we call a map $f : M \to N$ in $\mathcal{A}$ an *isomorphism mod $\mathcal{B}$* if Ker f and Cok f are in $\mathcal{B}$.

Define $\mathcal{A}/\mathcal{B}$ to be the category with the same objects as $\mathcal{A}$ but in which the arrows are obtained from the arrows in $\mathcal{A}$ by formally adjoining inverses for all isomorphisms mod $\mathcal{B}$.

In our particular case, any $M \to N$ in $\mathcal{A}/\mathcal{B}$ can be represented by a diagram

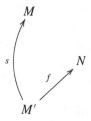

where s is an isomorphism mod $\mathcal{B}$ and $f : M' \to N$ is in $\mathcal{A}$, so we have

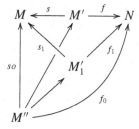

and $fs^{-1} = f_1 s_1^{-1}$ when there exists an M'' dominating the both of them.

Composition is given by

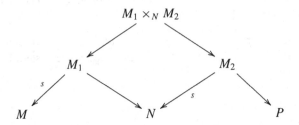

Fact $\mathcal{A}/\mathcal{B}$ is an abelian category. The functor $\mathcal{A} \to \mathcal{A}/\mathcal{B}$ is exact, and $\mathcal{B}$ is exactly the set of objects killed by this functor.

Suppose now V is in $\text{Modf}(S^{-1}A)$. Then $V = S^{-1}M$, where M is in Modf A, and consider

$$\text{Hom}_{S^{-1}A}(S^{-1}M, S^{-1}N).$$

Then we have

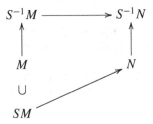

by "clearing denominators", whence

Modf $S^{-1}A \equiv$ Modf $A/(S - \text{torsion finitely generated modules})$.

Preview There is a basic exact sequence

$$K_1\,\mathcal{A}/\mathcal{B} \longrightarrow K_0\,\mathcal{B} \longrightarrow K_0\,\mathcal{A} \longrightarrow K_0\,\mathcal{A}/\mathcal{B} \longrightarrow 0.$$

Chapter 28
$K_0 A \cong K_0 A[T]$

Theorem *If A is a regular noetherian ring, then $K_0 A \cong K_0 A[T]$.*

Ingredients of Proof

1: Because A is regular noetherian, which implies that $A[T]$ is regular noetherian as well, we have $K_0 \mathcal{P}_A = K_0(\text{Modf } A)$, and moreover Modf A is an abelian category, and similarly for $A[T]$.

2: Localization Theorem: If $\mathcal{B}$ is a Serre subcategory of the abelian category $\mathcal{A}$, then we have the exact sequence

$$K_0 \mathcal{B} \longrightarrow K_0 \mathcal{A} \longrightarrow K_0(\mathcal{A}/\mathcal{B}) \longrightarrow 0 .$$

Put $\mathcal{B} = A[T_0, T_1]$, a ring graded by degree. Take $\mathcal{A}$ to be all finitely generated graded $\mathcal{B}$-modules, e.g., $M = \bigoplus_{n \geq 0} M_n$, so $\mathcal{B} = \bigoplus_{n \geq 0} \mathcal{B}_n$ where $\mathcal{B}_n = AT_0^n + \cdots + AT_1^n$ and check this is an abelian category by classical results. Consider

$$\mathcal{B}[T_0^{-1}] = A[T_1] \otimes_A A[T_0, T_0^{-1}] ,$$

where $A[T_0, T_0^{-1}]$ is the collection of Laurent polynomials. If $M = \bigoplus_{n \geq 0} M_n$, then $M[T_0^{-1}] = \tilde{M} \otimes_A A[T_0, T_0^{-1}] = \bigoplus_{n \geq 0} \tilde{M} T_0^n$, where $\tilde{M}$ is the degree zero part of $M[T_0^{-1}]$, i.e., $\tilde{M} = \varinjlim\{M_n \xrightarrow{\times T_0} M_{n+1}\}$.

3: Conclude that finitely generated $\mathbb{Z}$-graded modules over $\mathcal{B}[T_0^{-1}]$ may be identified with finitely generated modules over $A[T]$ where $T = T_1/T_0$.

We have $\mathcal{B} \subset \mathcal{A}$ with $\mathcal{B}$ the collection of finitely generated graded $\mathcal{B}$-modules killed by some power of T_0, so

$$\mathcal{B} \subset \mathcal{A} \longrightarrow \mathcal{A}/\mathcal{B}$$

© Springer Nature Switzerland AG 2020
R. Penner, *Topology and K-Theory*, Lecture Notes in Mathematics 2262,
https://doi.org/10.1007/978-3-030-43996-5_28

with $\mathcal{A}/\mathcal{B}$ the collection of finitely generated graded $B[T_0^{-1}]$-modules, i.e., $\mathrm{Modf}(A[T])$, which has the same K_0 as $A[T]$ by the Resolution Theorem.

We consider the projective modules in $\mathcal{A}$. A free graded B-module is a direct sum of various $B(p)$, where $B(p)$ is just B with the grading shifted p steps, i.e., $B(p)_n = B_{n-p}$.

Fact Any projective graded B-module P is of the form

$$P = B \otimes_A Q_0 \oplus B \otimes_A Q_1[1] \oplus \cdots,$$

where $Q_0, Q_1, \ldots$ are projective A-modules and

$$B \otimes_A Q_1[1] = B(1) \otimes_A Q_1.$$

In general, define an increasing filtration $F_p M$ given by the B-submodule generated by $M_0, M_1, \ldots, M_p$, so we have

$$B(p) \otimes (M_p/B_1 M_{p-1} + \cdots + B_p M_0)$$
$$\downarrow$$
$$F_p M / F_{p-1} M$$

These are clearly isomorphisms for free modules, which implies also isomorphisms for projectives since projective implies summand of free and the construction is functorial.

Thus

$$K_0 \begin{pmatrix} \text{graded finitely generated} \\ \text{projective } B \text{-modules} \end{pmatrix} = K_0(\mathcal{P}_A)[\xi]$$

where

$$B(n) \otimes_A Q \longleftarrow [Q]\xi^n$$
$$[P] \longmapsto \sum [Q_n]\xi^n$$

with $Q_n = (P/B^+ P)_n$ and $B^+ = B_1 + B_2 + \cdots$ the augmentation ideal.

The Szyzygy Theorem implies that any $M \in \mathcal{A}$ has a finite resolution by graded finitely generated projective B-modules whence

$$K_0 \mathcal{A} = K_0(A)[\xi].$$

The Devissage Theorem implies that $K_0 \mathcal{B} = K_0 \begin{pmatrix} \text{finitely generated graded} \\ \text{modules over } B/T_0 B \end{pmatrix}.$

But $A[T_1] = B/T_0 B$, and by the argument above

$$K_0 B = K_0(A)[\xi]$$

$$Q \otimes_A B/T_0 B(n) \longleftarrow [Q]\xi^n .$$

We thus have

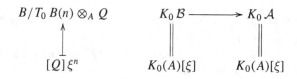

and also have a resolution of, for instance, $B/T_0 B \otimes Q$ given by

$$0 \longrightarrow B(1) \otimes_A Q \xrightarrow{T_0} B \otimes Q \longrightarrow B/T_0 B \otimes Q \longrightarrow 0 ,$$

so in $K_0 \mathcal{A}$, $B/T_0 B(n) \otimes Q$ maps to

$$[Q]\xi^n - [Q]\xi^{n+1} .$$

It follows that the map $K_0 B \to K_0 \mathcal{A}$ is multiplication by $(1 - \xi)$, whose cokernel is $K_0 A$, as desired.   □

Note that map is

$$B \rightsquigarrow B \oplus_A Q \rightsquigarrow (B \otimes_A Q)[T_0^{-1}] \rightsquigarrow B[T_0^{-1}] \otimes_A Q$$

and take the degree zero part.

[Paul Goerss points out that this map is actually induced by augmentation.]

K_0 is functorial with respect to exact functors.

Proof of 2 We have

$$\mathcal{B} \subset \mathcal{A} \xrightarrow{\gamma} \mathcal{B}/\mathcal{A}$$

and think of the examples S-torsion finitely generated A modules $\subset$ finitely generated A modules $\to$ finitely generated $S^{-1}A$ modules.

Given any object V of $\mathcal{A}/\mathcal{B}$, choose $M \in \mathcal{A}$ with $\gamma M \cong V$. The point is such an M is unique up to isomorphism mod B, i.e.,

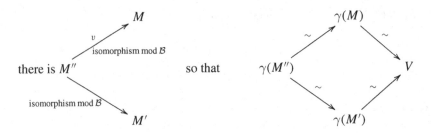

where $[M] - [M''] = [\operatorname{Cok} u] - [\operatorname{Ker} u]$ in $K_0 A$, so $[M] \cong [M''] \cong [M']$ in $K_0 \mathcal{A}/K_0 \mathcal{B}$ because $[\operatorname{Cok} u]$ and $[\operatorname{Ker} u]$ lie in $K_0 \mathcal{B}$ since $\mathcal{B} \subset \mathcal{A}$ is Serre.

We therefore associate to V an element of $K_0 \mathcal{A}/K_0 \mathcal{B}$. This is additive for exact sequences hence is a homomorphism

$$K_0 \mathcal{A}/K_0 \mathcal{B} \xleftarrow{\quad} K_0(\mathcal{A}/\mathcal{B}) .$$

Easy to check these are inverses.   □

Another Example Take G to be a finite group and $\mathbb{Q}_p$ to be the p-adic rationals and $\mathbb{Z}_p$ the p-adic integers. Let $R(G, \mathbb{Q}_p)$, $R(G, \mathbb{Z}/p\,\mathbb{Z})$ and $R(G, \mathbb{Z}_p)$ be the respective Grothendieck groups of representations of G over $\mathbb{Q}_p$, $\mathbb{Z}/p\,\mathbb{Z}$ and $\mathbb{Z}_p$.

We have

$$\{\text{finitely generated torsion } \mathbb{Z}_p[G] - \text{modules}\} \longleftrightarrow \operatorname{Modf}(\mathbb{Z}_p[G]) \twoheadrightarrow \operatorname{Modf}(\mathbb{Q}_p[G])$$

and get

$$
\begin{array}{ccccc}
K_0(\text{torsion}) & \longrightarrow & K_0(\operatorname{Modf} \mathbb{Z}_p[G]) & \longrightarrow & R(G, \mathbb{Q}_p) \longrightarrow 0 \\
\| \text{ by Devissage} & & \| \text{ by Resolution theorem} & & \\
R(G, \mathbb{Z}/p\,\mathbb{Z}) & & R(G, \mathbb{Z}_p) & &
\end{array}
$$

i.e., we have

$$R(G, \mathbb{Z}/p\,\mathbb{Z}) \xrightarrow{\ \alpha\ } R(G, \mathbb{Z}_p) \longrightarrow R(G, \mathbb{Q}_p) \longrightarrow 0$$

$$\rho \downarrow \quad \begin{array}{c} M \\ \updownarrow \\ M/PM \end{array} \quad \text{exact functor for } M \text{ free over } \mathbb{Z}_p$$

$$R(G, \mathbb{Z}/p\,\mathbb{Z})$$

where α takes a representation W over $\mathbb{Z}/p\,\mathbb{Z}$ and sends $[W]$ to $[P_0] - [P_1]$, where

$$0 \longrightarrow P_1 \longrightarrow P_0 \longrightarrow W \longrightarrow 0$$

is an exact sequence of G-modules and P_0 and P_1 are free over $\mathbb{Z}_p$. Thus

$$\rho \alpha [\omega] = [P_0/p P_0] - [P_1/p P_1].$$

But we have

$$\mathrm{Tor}_1^{\mathbb{Z}_p}(\mathbb{Z}/p\,\mathbb{Z}, W) \longrightarrow P_1/p P_1 \longrightarrow P_0/p\,\mathbb{Z} \longrightarrow W \longrightarrow 0$$

$$\Big\| \\ W$$

Thus

$$\rho \alpha [W] = [W] - [W] = 0,$$

and we therefore get an induced map $R(G, \mathbb{Q}_p) \to R(G, \mathbb{Z}/p\,\mathbb{Z})$ called the *specialization mod p*. There is also a lifting map $R(G, \mathbb{Z}/p\,\mathbb{Z}) \to R(G, \mathbb{Q}_p)$ which is its inverse. See Serre's book.

Chapter 29
Classifying Spaces

Geometric Realization of a Simplicial Space

Recall that Δ is the category of posets $[p] = \{0, \cdots, p\}$ in the usual order and maps of posets. We have the functor

$$\Delta \longrightarrow \text{Top} = \text{the category of topological spaces and continuous maps}$$
$$[p] \longmapsto \Delta_p = \text{simplex with vertices } \{0, \cdots, p\}.$$

A *(semi-)simplical* space (s.s.) X is a functor

$$\Delta^{\text{op}} \longrightarrow \text{Top}$$
$$[p] \longmapsto X_p.$$

The *geometric realization* $|X|$ of X is defined as a coequalizer

$$\coprod_{[p] \to [q]} X_q \times \Delta_p \rightrightarrows \coprod_{[p]} X_p \times \Delta_p \to |X|.$$

Note that $\text{Hom}_{\text{Top}}(|X|, Y) = \text{Hom}_{\text{s.s.}}(X, [p] \mapsto Y^{\Delta_p})$.

Classifying Space of a Topological Category $\mathcal{C}$

We have the diagram

$$\cdots \text{Ar}\,\mathcal{C} \underset{\text{Ob}\,\mathcal{C}}{\times} \text{Ar}\,\mathcal{C} \rightrightarrows \text{Ar}\,\mathcal{C} \rightrightarrows \text{Ob}\,\mathcal{C}.$$

© Springer Nature Switzerland AG 2020
R. Penner, *Topology and K-Theory*, Lecture Notes in Mathematics 2262,
https://doi.org/10.1007/978-3-030-43996-5_29

In general in a category $\mathcal{T}$ where we have fibered products, we have also the notion of a *category object* $\mathcal{C}$. It is $\mathrm{Ob}\,\mathcal{C}$ and $\mathrm{Ar}\,\mathcal{C}$ plus maps

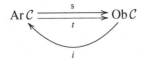

and

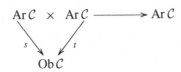

so that for any $Y \in \mathcal{T}$, $\mathrm{Hom}_{\mathcal{T}}(Y, \cdot)$ carries this data into a category.

A *topological category* is a category object $\mathcal{C}$ in Top. That is, the set of objects $\mathrm{Ob}(\mathcal{C})$ and the set of morphisms $\mathrm{Ar}(\mathcal{C})$ are made into topological spaces so that usual maps such as the composition $\mathrm{Ar} \times_{\mathrm{Ob}} \mathrm{Ar} \to \mathrm{Ar}$, are continuous.

The *nerve* of a topological category $\mathcal{C}$ is the simplicial space

$$\ldots \mathrm{Ar} \underset{\mathrm{Ob}}{\times} \mathrm{Ar} \underset{\mathrm{Ob}}{\times} \mathrm{Ar} \Longrightarrow \mathrm{Ar} \underset{\mathrm{Ob}}{\times} \mathrm{Ar} \Longrightarrow \mathrm{Ar}\,\mathcal{C} \longrightarrow \mathrm{Ob}\,\mathcal{C},$$

and the *classifying space* is the geometric realization $|\text{nerve of } \mathcal{C}|$ of the nerve.

Example Take G a topological group. The nerve is

$$\ldots G \times G \times G \Longrightarrow G \times G \overset{\pi_2}{\underset{\pi_1}{\longrightarrow}} G \Longrightarrow *$$

$$G \times G \times G \Longrightarrow G \times G \Longrightarrow G \quad \equiv \quad \text{nerve of cofibered category defined by } \mathrm{Hom}(*, \cdot)$$

with vertical maps pr_{12}, pr_1 down to

$$G \times G \Longrightarrow G \Longrightarrow *$$

Taking $PG = |\text{top line}|$, $BG = |\text{bottom line}|$, then is a principal fiber bundle over BG with structure group G. Moreover, it is universal in the sense that $[X, BG]$ is the collection of isomorphism classes of principal G-bundles over X for X paracompact.

Example Take a poset I. Then we can see that BI is a simplical complex whose simplexes are chains in I.

Fun Example Fix a vector space $V = \mathbb{C}^n$ over $\mathbb{C}$ and consider the poset of proper subspaces of V. This is topological poset so a topological category. Consider a self-adjoint operator A on V with $0 \le A \le I$. Then it has eigenspaces, so if the eigenvalues are $0 \le \lambda_1 < \cdots < \lambda_p \le I$, then we get a flag $0 \subset V_1 \subset V_2 \subset \cdots \subset V_p \subset V$. We get an identification of this classifying space with a certain family of self-adjoint operators. **Exercise** to fix this: try all rays of A's with tr $A = 0$. If this works, then $B\mathcal{C} = S^{\frac{n(n+1)}{2}-2}$, a sphere.

Suppose $\mathcal{C}$ is a discrete category. What are $\pi_0 B\mathcal{C}$, $\pi_1(B\mathcal{C}, *)$, $H_*(B\mathcal{C}, \mathbb{Z})$? It is clear that $\pi_0 B\mathcal{C}$ is the set of components of $\mathcal{C}$. Also, for $X \in \mathrm{Ob}\,\mathcal{C}$, what is $\pi_1(B\mathcal{C}, X)$? The fundamental groupoid of $B\mathcal{C}$ is equivalent to the groupoid obtained from $\mathcal{C}$ by inverting the arrows of $\mathcal{C}$ as for covering spaces. In effect given a covering space E of $B\mathcal{C}$, we get a functor from $\mathcal{C}$ to Sets by

$$X \longmapsto \text{fiber of } E \text{ over } X$$
$$f \longmapsto \text{deck transformation.}$$

This carries maps to isomorphisms.

Conversely, given $F : \mathcal{C} \to$ Sets so that $F(f)$ is an isomorphism for all f, we get a covering space by taking $\mathcal{C}_F$ to be the cofibered category over $\mathcal{C}$ defined by the functor F, so $\mathrm{Ob}\,\mathcal{C}_F = \{(X, \xi) : \xi \in F(X)\}$, and finally forming $B\mathcal{C}_F$.

Chapter 30
Higher K-Groups

We saw that $\pi_0 BC$ is the set of components of C in the usual sense. If we think of objects X of C as being points in BC, then we can speak of $\pi_1(BC, X)$. Now the fundamental groupoid of BC is simply the groupoid obtained from C by inverting all the arrows. For given a covering space E of BC, we get a functor $C \to$ Sets

$$
\begin{array}{ccc}
X & \longmapsto & \text{fiber of } C \text{ over } X \\
\downarrow {\scriptstyle f} & & \downarrow \cong \text{ lifting of path association to } f \\
Y & \longmapsto & \text{fiber of } C \text{ over } Y
\end{array}
$$

which inverts morphisms. Conversely, given $F : C \to$ Sets so that $F(f)$ is an isomorphism for all f, we get a covering space by taking C_F to be the cofibered category over C defined by F and forming $E = BC_F$. So $\pi_1(BC, X)$ is just the group of "loops" based at X in the groupoid obtained from C by inverting arrows.

Fun Example of Chapter 29 Fixed Consider the topological poset of all subspaces of a finite-dimensional vector space V. The nerve is

$$
\cdots \Longrightarrow \coprod_{p \leq q} \mathrm{Flag}_{p,q}(V) \Longrightarrow \coprod_{p=0}^{n} \mathrm{Flag}_{p}(V) .
$$

Then the realization of this s.s. is the space of self-adjoint operators A on V so that $0 \leq A \leq I$. Call this space Y. To each subspace W we associate the projection operator p_W, and to an inclusion $W' \subset W$ we associate the "1-simplex"

$$
t_0 \, p_{W'} + t_1 \, p_W
$$

© Springer Nature Switzerland AG 2020
R. Penner, *Topology and K-Theory*, Lecture Notes in Mathematics 2262,
https://doi.org/10.1007/978-3-030-43996-5_30

for $t_0 + t_1 = 1, 0 \leq t_0, t_1 \leq 1$. Similarly for higher dimensions, so that a point in the realization of the form

$$(t_0, t_1, \cdots, t_p) \times (W_0 < W_1 < \cdots < W_p)$$

for $(t_0, \cdots, t_p) \in \text{Int } \Delta_p$, and $W_0 < W_1 < \cdots < W_p$ a flag of length p, is mapped to the self-adjoint operator

$$\sum_{j=0}^{p} t_j \, p_{W_j} \, .$$

Check that this works.

Now for a small category $\mathcal{C}$, we have seen that $B\mathcal{C} = |\text{nerve of } \mathcal{C}|$ has fundamental groupoid equivalent to the groupoid obtained by inverting the arrows of $\mathcal{C}$.

Example If $\mathcal{C}$ is a poset, then $B\mathcal{C}$ is the simplicial complex of chains in $\mathcal{C}$. Since $B\mathcal{C}$ is a CW-complex, the homology of $B\mathcal{C}$ with coefficients in a local coefficient system L can be computed cellularly

$$C_q(B\mathcal{C}, L) = \bigoplus_{\substack{X_0 \leftarrow \cdots \leftarrow X_q \\ \text{none the identity map}}} L(X_q) \, .$$

On the other hand, we have $H_q(\mathcal{C}, L)$, where L is viewed as a covariant functor $\mathcal{C} \to \text{Ab}$. This homology is computed via the simplicial abelian group $C(\mathcal{C}, L)$

$$\cdots \Rrightarrow \bigoplus_{X_2 \to X_1 \to X_0} L(X_2) \Rightarrow \bigoplus_{X_1 \to X_0} L(X_1) \Rightarrow \bigoplus_{X_0} L(X_0)$$

which is the same as $C(B\mathcal{C}, L)$ except for degeneracies. Thus, the Normalization Theorem says

$$H_*(C_+(\mathcal{C}, L)) = H_*(C_*(\mathcal{C}, L)/\text{degeneracies})$$
$$\cong H_*(C_*(B\mathcal{C}, L)) \, ,$$

so we have an isomorphism

$$H_*(B\mathcal{C}, L) \cong H_*(\mathcal{C}, L)$$

between "geometric" and "derived-functor" homologies.

Now we recall the Whitehead theorem. If $f : X \to Y$ is a map of CW complexes, then f is a homotopy equivalence if and only if f induces isomorphisms

$$\pi_0(X) \xrightarrow{\sim} \pi_0(Y) \, ,$$

$$\pi_1(X, x) \xrightarrow{\sim} \pi_1(Y, f(x)), \text{ for all } x \in X,$$

$H_q(X, f^{-1}(L)) \xrightarrow{\sim} H_q(Y, L)$, for all q and all local coefficient systems L on Y. Therefore, a functor $f : C \to C'$ induces a homotopy equivalence $BC \to BC'$ when f induces an isomorphism on associated groupoids and homology with coefficients in any morphism-inverting Ab-valued functor.

Fun Example of Chapter 29 Again Let C be the poset of subspaces of V. Consider

$$\left| \begin{array}{l} \text{s.s. of } U \subset W_0 < W_1 < \cdots < W_r \subset V \\ \text{so that either } W_0 > 0 \text{ or } W_1 < V \end{array} \right| \subset BC$$

The space on the left can be written

$$B(\text{all subspaces } W > U) \bigcup_{B(\text{proper subspaces})} B(\text{all subspaces } W < V).$$

Now

$$Y = \left\{ \text{self-adjoint } A \mid 0 \le A \le I \right\}$$

is homeomorphic to $D^{n(n+1)/2}$, so $\partial Y \cong S^{n(n+1)/2-1}$. If we set

$$Z = B(\text{proper subspaces})$$

then

$$\left| \begin{array}{l} 0 \subset W_0 < \cdots < W_p \subset V \text{ so} \\ \text{that } W_0 > 0 \text{ or } W_p < V \end{array} \right| = \partial Y = CZ \bigcup_Z CZ,$$

so

$$SZ = \partial Y \cong S^{n(n+1)/2-1},$$

and with more work we can show $Z \cong S^{n(n+1)/2-2}$.

Definition of Higher K-groups

Let $\mathcal{M}$ be an exact category. Define $Q\mathcal{M}$ to be the category with the same objects as $\mathcal{M}$, but in which an arrow $M' \to M$ is an isomorphism of M' with an admissible subquotient of M, where an admissible subquotient of M is something of the form M_1/M_2, where $0 \subset M_2 \subset M_1 \subset M$ is an admissible filtration of M. So a map from M' to M in $Q\mathcal{M}$ is an isomorphism

$$M' \xrightarrow{\sim} M_1/M_2.$$

Composition is clear.

The *higher K-groups* of an exact category $\mathcal{M}$ are defined by

$$K_i \mathcal{M} = \pi_{i+i}(BQ\mathcal{M}, 0).$$

For this definition to be reasonable, it should correspond with the Grothendieck definition for $i = 0$. First we show $\pi_1(BQ\mathcal{M})$ is abelian. Direct sum gives a functor

$$Q\mathcal{M} \times Q\mathcal{M} \longrightarrow Q\mathcal{M},$$

so we get a map

$$B(Q\mathcal{M} \times Q\mathcal{M}) \longrightarrow BQ\mathcal{M}$$
$$\Big\|$$
$$BQ\mathcal{M} \times BQ\mathcal{M}$$

providing an operation on $BQ\mathcal{M}$. It follows that $BQ\mathcal{M}$ is an H-space, and this implies that $\pi_1(BQ\mathcal{M})$ is abelian.

Picture: In $Q\mathcal{M}$, we have maps

$$0 \xrightarrow[\text{0 sub-object}]{\text{0 quotient object}} M$$

For any object M, this gives a loop in $BQ\mathcal{M}$ associated to M. For an exact sequence

$$0 \longrightarrow M' \longrightarrow M \longrightarrow M'' \longrightarrow 0$$

in $\mathcal{M}$ we have arrows

$$0 \underset{0=0/0}{\overset{\substack{0=M'/M \\ 0=M'/M'}}{\rightrightarrows}} M$$

in $Q\mathcal{M}$. Then we get a homotopy of the loop associated to M' with one of those associated to M via

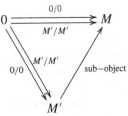

But we also have

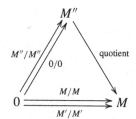

Hence the loop $0 \xrightarrow[0/0]{M/M} M$ is the loop of M' followed by that of M''.

Chapter 31
The Category $Q\mathcal{M}$

Suppose that $\mathcal{M}$ is an exact category and let $Q\mathcal{M}$ be the category with the same objects but in which a morphism $M' \to M$ is an isomorphism of M' with an admissible subquotient of M, i.e., $0 \subset M_0 \subset M_1 \subset M$ with all quotients in $\mathcal{M}$ and M_1/M_0 is a sub-quotient.

An *admissible monomorphism* $i : M' \hookrightarrow M$ is an injection so that $M/M' \in \mathcal{M}$. Such determines a map in $Q\mathcal{M}$ to be denoted

$$i_! : M' --- \to M \, .$$

Likewise, an *admissible epimorphism* $j : M \twoheadrightarrow M''$ in $\mathcal{M}$ determines a map

$$j^! : M'' --- \to M$$

in $Q\mathcal{M}$.

Suppose given an arbitrary map $N --- \to M$ in $Q\mathcal{M}$ determined by $\theta : N \xrightarrow{\sim} M_1/M_0$, where $0 \subset M_0 \subset M_1 \subset M$ is an admissible filtration. We get a diagram

$$
\begin{array}{ccc}
M_1 & \overset{i'}{\lhook\joinrel\longrightarrow} & M \\
{}^{\text{``}\theta^{-1}\text{''}}\big\downarrow{}^{j'} & & \big\downarrow{}^{j} \\
N & \underset{\underset{\text{``}\theta\text{''}}{i}}{\lhook\joinrel\longrightarrow} & M/M_0
\end{array}
$$

in $\mathcal{M}$. The original $N --- \to M$ in $Q\mathcal{M}$ is either of

© Springer Nature Switzerland AG 2020

R. Penner, *Topology and K-Theory*, Lecture Notes in Mathematics 2262,
https://doi.org/10.1007/978-3-030-43996-5_31

$$
\begin{array}{ccc}
M_1 & \xrightarrow{\ (i')_!\ } & M \\
\Big\uparrow {\scriptstyle (j')^!} & & \Big\uparrow {\scriptstyle j^!} \\
N & \xrightarrow[\ i_!\ ]{} & M/M_0
\end{array}
$$

Proposition *A functor $Q\mathcal{M} \xrightarrow{F} \mathcal{C}$ is the "same thing" as giving for each object $M \in \mathcal{M}$ an object $F(M)$ in $\mathcal{C}$, for each admissible monomorphism $i : N \hookrightarrow M$ a map $i_* : F(N) \to F(M)$ and for each admissible epimorphism $j : M \to M''$ a map $j^* : F(M'') \to F(M)$ so that*

1: $(\mathrm{id}_M)_* = \mathrm{id}_{F(M)}$ *and* $(i_1\, i_2)_* = i_{1*}\, i_{2*}$,

2: $(\mathrm{id}_M)^* = \mathrm{id}_{F(M)}$ *and* $(j_1\, j_2)^* = j_2^*\, j_1^*$,

3: for any cartesian square

$$
\begin{array}{ccc}
M & \overset{i}{\lhook\joinrel\longrightarrow} & N \\
{\scriptstyle j}\Big\downarrow & & \Big\downarrow{\scriptstyle j'} \\
M' & \underset{i'}{\lhook\joinrel\longrightarrow} & N'
\end{array}
$$

of admissible monomorphisms and epimorphisms, we have

$$
i_*\, j^* = (j')^*\, i'_* .
$$

Proof Exercise ... except for the following remark. We can think of a map $M-\,-\,-\,-\to N$ in $Q\mathcal{M}$ as being an isomorphism class of diagrams

$$
\begin{array}{ccc}
N_1 & \lhook\joinrel\longrightarrow & N \\
\Big\downarrow & & \\
M & &
\end{array}
$$

and composition of such given by

$$
\begin{array}{ccccc}
N_1 \times_N P_1 & \lhook\joinrel\longrightarrow & P_1 & \lhook\joinrel\longrightarrow & P \\
\Big\downarrow & & \Big\downarrow & & \\
M_1 & \lhook\joinrel\longrightarrow & N & & \\
\Big\downarrow & & & & \\
M & & & \square &
\end{array}
$$

Theorem $\pi_1(BQ\mathcal{M}, 0) \cong K_0\, \mathcal{M}$.

Proof The idea is to look at morphism-inverting functors F from $Q\mathcal{M}$ to Sets, which are essentially the same as covering spaces of $BQ\mathcal{M}$ and to show that $K_0\mathcal{M}$ acts naturally on $F(0)$, and that in this way the category of such F becomes equivalent to $(K_0\mathcal{M})$-sets. To each M we have $i_M : 0 \hookrightarrow M$ and

$$(i_M)_! : 0 - - - - \to M \quad \text{in} \quad Q\mathcal{M}$$

and

$$(i_M)_* = F(i_{M!}) : F(0) \xrightarrow{\sim} F(M).$$

Replace F by a map so that these are the identity; this is like choosing a maximal tree. Thus we have

$$F_{(i_!)} = \mathrm{id}_{F(0)}, \quad \text{for all} \quad i : M \hookrightarrow N$$

and

$$\begin{array}{ccc}
\mathrm{Ker}\, j \hookrightarrow & M_1 \stackrel{i}{\hookrightarrow} & M \\
\downarrow & \downarrow j & \\
0 \hookrightarrow & N & \\
& i_N &
\end{array}$$

Apply F to see that $F(j^!)$ depends only on $\mathrm{Ker}\, j$. To each N, let $P_N : N \twoheadrightarrow 0$. Then to each N we get an isomorphism $F(P_N^!) : F(0) \to F(0)$, and we have thus shown that

$$F(j^!) = F(P_{\mathrm{Ker}\, j}^!),$$

and in fact

$$F \left(\begin{array}{c} M \\ N \qquad M \end{array} \right) = F(P_{\mathrm{Ker}\, j}^!)$$

Given $0 \to N' \to N \to N'' \to 0$ we have

$$\begin{array}{ccc}
N' \hookrightarrow & N \\
\downarrow & \downarrow \\
0 \hookrightarrow & N'' \\
& \downarrow P_{N''} \\
& 0
\end{array}$$

so

$$F(P_N^!) = F(P_{N'}^!) \circ F(P_{N''}^!).$$

So to each N in $\mathcal{M}$ we have associated an automorphism $F(P_N^!)$ of $F(0)$.

But $K_0\mathcal{M}$ is the not necessarily abelian group generated by $[N]$ for $N \in \mathrm{Ob}\,\mathcal{M}$ with the relations $[N] = [N'][N'']$ for each exact $0 \to N' \to N \to N'' \to 0$.

Note however that

$$[N'][N''] = [N' \oplus N''] = [N'' \oplus N'] = [N''][N'],$$

so we get by universal nonsense a map

$$K_0\mathcal{M} \longrightarrow \mathrm{Aut}(F(0))$$
$$[N] \longmapsto F(P_N^!).$$

Thus $F(0)$ is a $(K_0\mathcal{M})$-set.

Conversely, given a $(K_0\mathcal{M})$-set S, define

$$F_S : Q\mathcal{M} \longrightarrow \mathrm{Sets}$$

by $F_S(M) = S$ and $F_S(i_!) = \mathrm{id}_S$ with $F_S(j^!)$ given by multiplication by $[\mathrm{Ker}\,j] \in K_0\mathcal{M}$.   □

Chapter 32
Homotopy Equivalence

A functor $F : C \to C'$ is a *homotopy equivalence* if the induced map

$$BC \longrightarrow BC'$$

is a homotopy equivalence. Call C *contractible* if $BC \cong *$, i.e., if BC is homotopy equivalent to a point.

(Nice) **Theorem** $B(C \times C') \to BC \times BC'$ is a homeomorphism for $\times$ on the right the CW-product of CW complexes and $\times$ on the left the category product.

Proof Follows from Milnor's theorem that $|X \times Y| = |X| \times |Y|$ for X, Y simplicial sets. $\qquad\square$

Take in particular C' to be the poset $\{0 < 1\}$ so that $BC' = [0, 1]$. Therefore

$$B(C \times \{0 < 1\}) \cong BC \times [0, 1].$$

A functor $F : C \times \{0 < 1\} \to C'$ is the same as a pair of functors $f_0, f_1 : C \to C'$ and a natural transformation F from f_0 to f_1. Thus

$$BC \times [0, 1] \cong B(C \times \{0, 1\}) \xrightarrow{B(F)} BC'$$

is a homotopy between $B(f_0)$ and $B(f_1)$. Therefore two functors $f, g : C \to C'$ which can be connected by a finite sequence (zigzag) of natural transformations

$$f = f_0 \longrightarrow f_1 \longleftarrow f_2 \longrightarrow \cdots \longleftarrow f_n = g$$

induce homotopic maps $Bf \simeq Bg : BC \to BC'$.

© Springer Nature Switzerland AG 2020

R. Penner, *Topology and K-Theory*, Lecture Notes in Mathematics 2262,
https://doi.org/10.1007/978-3-030-43996-5_32

Example 32.1 If $f : C \to C'$ has an adjoint $g : C' \to C$, then we have adjunction maps

$$gf \longleftarrow \mathrm{id} \qquad fg \longrightarrow \mathrm{id}$$

so f and g are homotopy equivalences.

Example 32.2 A category with final or initial object is automatically contractible. In particular, this holds for any abelian category.

Example 32.3 If a category C has an object x_0, then we have the constant functor $\underline{x_0} : C \to C$ sending every object to x_0 and every morphism to the identity of x_0. Suppose, in addition, that we have a functor $F : C \to C$ and natural transformations

$$\underline{x_0} \longleftarrow F \longrightarrow \mathrm{Id}_C .$$

Then C is contractible.

A useful condition for a functor to be a homotopy equivalence is

Theorem A $f : C \to C'$ *is a homotopy equivalence when f/Y is contractible for all $Y \in \mathrm{Ob}\, C'$ or when Y/f is contractible for all $Y \in \mathrm{Ob}\, C'$.*

Proof From last term, either condition is sufficient to prove homology equivalence, and the isomorphism of π_1 is easy when π_1 is viewed categorically. The result then follows from Whitehead's Theorem. $\qquad\square$

Recall that f/Y has objects (X, u)

$$u : fX \longrightarrow Y .$$

Devissage Theorem *Suppose that A is an abelian category and B is a full sub-category closed under sub-objects, quotient objects and products. (Think of $A = \mathrm{Modf}\,(A)$, $B = \mathrm{Modf}\,(A/I)$.) Assume every $M \in \mathrm{Ob}\, A$ has a finite filtration whose quotients are in B (e.g., $I^n = 0$). Then the inclusion*

$$QB \subset QA$$

is a homotopy equivalence, so B and A have the same K-groups.

Proof Let $f : QB \hookrightarrow QA$. Let $M \in \mathrm{Ob}\, QA = \mathrm{Ob}\, A$. What is f/M? It consists of (N, u), $N \in \mathrm{Ob}\, B$ and $u : N ---\to M$ in QA. Easy to see f/M is equivalent to the poset J_M of all *layers* (M_0, M_1) in M, i.e., $M_0 \subset M_1 \subset M$ so that $M_1/M_0 \subset \mathrm{Ob}\, B$, and where the ordering in J_M is

$$(M_0 \subset M_1) \leq (M_0' \subset M_1')$$

if and only if

$$M_0' \subseteq M_0 \subseteq M_1 \subseteq M_1' .$$

We know there exists

$$0 \subset F_1 \subset \cdots \subset F_p = M$$

so that $F_p / F_{p-1} \in \operatorname{Ob} \mathcal{B}$. We shall show that each inclusion in

$$J_0 \subset J_{F_1} \subset J_{F_2} \subset \cdots \subset J_{F_p} = J_M$$

is a homotopy equivalence, and it is enough to show that

$$J_{M'} \subset J_M \text{ is a homotopy equivalence when } M'/M \in \operatorname{Ob} \mathcal{B} .$$

Now,

$$(*) \qquad J_M \ni (M_0, M_1) \leq (M' \cap M_0, M_1) \geq (M' \cap M_0, M' \cap M_1) \in J_{M'}$$

and $(M' \cap M_0, M_1)$ is in J_M since $M_1/(M' \cap M_0) \subset (M/M') \times (M_1/M_0)$ with both M/M' and M_1/M_0 in $\mathcal{B}$.

Thus we have

$$i : J_{M'} \longrightarrow J_M , \quad \text{inclusion,}$$
$$r : (M_0, M_1) \longmapsto (M' \cap M_0, M' \cap M_1)$$

and $ri = \operatorname{id}$ while $(*)$ gives $\operatorname{id} \simeq ir$.

Now use Theorem A. $\qquad\qquad\qquad\qquad\qquad\qquad\qquad\qquad\qquad\qquad\qquad$ □

(Special case of) **Resolution Theorem** Suppose $\mathcal{M}$ is an exact category and $\mathcal{P}$ is a full subcategory closed under extensions in $\mathcal{M}$. Assume

(i) for $0 \to M' \to P \to M \to 0$ exact in $\mathcal{M}$; $P \in \mathcal{P}$ implies $M' \in \mathcal{P}$,
(ii) given $M \in \mathcal{M}$, there is an exact sequence

$$0 \longrightarrow P_1 \longrightarrow P_0 \longrightarrow M \longrightarrow 0 \quad \text{with } P_0, P_1 \in \mathcal{P} .$$

Then $Q\mathcal{P} \hookrightarrow Q\mathcal{M}$ is a homotopy equivalence.

Proof Factor into

$$Q\mathcal{P} \xrightarrow{g} \mathcal{C} \xrightarrow{f} Q\mathcal{M}$$

where $\mathcal{C}$ is the full subcategory of $Q\mathcal{M}$ whose objects are the P in $\mathcal{P}$. Note that in $\mathcal{C}$ a morphism $P' \dashrightarrow P$ is given by an isomorphism $P' \simeq M_1/M_0$ where $0 \subset M_0 \subset M_1 \subset P$ is admissible in $\mathcal{M}$.

We consider g/P for $P \in \mathcal{P}$. g/P is equivalent to the poset of $\mathcal{M}$-admissible layers (M_0, M_1) in P with $M_1/M_0 \in \mathcal{P}$, where we have

$$(M_0', M_1') \leq (M_0, M_1)$$

if and only if $M_0 \subseteq M_0' \subseteq M_1' \subseteq M_1$ and all quotients of this are in $\mathcal{P}$.

This poset is contractible since

$$(M_0, M_1) \leq (0, M_1) \geq (0, 0) ,$$

for which we need property (i). Thus by Theorem A, g is a homotopy equivalence.

We consider M/f for $M \in \mathcal{M}$. An object of this category is a P in $\mathcal{C}$ and a map $M \overset{u}{\dashrightarrow} P$ in $Q\mathcal{M}$.

First step: Introduce the full subcategory $\mathcal{F}$ of all (P, u) where u comes from an epimorphism. Show

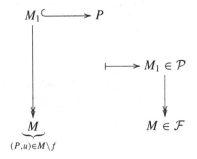

is adjoint to inclusion of $\mathcal{F} \subset M \backslash f$.

Second step: $\mathcal{F}^{\mathrm{op}}$ consists of all $\begin{smallmatrix} P \\ \downarrow \\ M \end{smallmatrix}$, for $P \in \mathcal{P}$, and morphisms are

This category has a "conical contraction" as in Example 32.3, namely fix

$$P_0 \quad \text{so we have} \quad P_0 \xleftarrow{\ \text{pr}_1\ } P_0 \times_M P \xrightarrow{\ \text{pr}_2\ } P$$

$$\begin{array}{cccc}
P_0 & P_0 & P_0 \times_M P & P \\
\downarrow & \downarrow & \downarrow & \downarrow \\
M & M & M & M
\end{array}$$

Thus $M \backslash f \cong *$, so by Theorem A, f is a homotopy equivalence, as desired. $\square$

Chapter 33
A Filtration of $Q(\mathcal{P}_A)$

Let A be a Dedekind domain and F its field of fractions. Consider $Q(\mathcal{P}_A)$. There is a natural filtration

$$F_n \, Q(\mathcal{P}_A) = \text{ full subcategory consisting of those } P \in Q(\mathcal{P}_A)$$
$$\text{of rank at most } n.$$

Since we work over a Dedekind domain, two $P, Q \in \mathcal{P}_A$ are isomorphic when they have the same rank r, and in this case

$$\Lambda^r P \cong \Lambda^r Q \in \text{Pic } A \,.$$

Thus,

$$\{0\} = F_0 \, Q(\mathcal{P}_A) \subset F_1 \, Q(\mathcal{P}_A) \subset \cdots ,$$

and $F_n \, Q(\mathcal{P}_A)$ is obtained from $F_{n-1} \, Q(\mathcal{P}_A)$ by adding one new isomorphism class for each element of $\text{Pic}(A)$.

Let us "compute" $H_*(F_n \, Q(\mathcal{P}_A), F_{n-1} \, Q(\mathcal{P}_A))$.

To this end, fix $P \in \mathcal{P}_A$ and consider $F_{n-1} \, Q(\mathcal{P}_A)/P$. This is equivalent to the poset of admissible layers (P_0, P_1) in P where rank $(P_1/P_0) < n$ with the ordering $(P_0', P_1') \leq (P_0, P_1)$ if $P_0 \subseteq P_0' \subseteq P_1' \subseteq P_1$.

First observation: Admissible sub-objects $P' \subset P$ are in one-to-one correspondence with subspaces V' of $V = F \otimes_A P$ via

$$F \otimes_A P' = V' \quad \text{where} \quad P' = V' \cap P \subset V \,.$$

Let V be a vector space over F and let $Y(V)$ be the poset of layers (V_0, V_1) in V such that $V_0 \neq 0$ or $V_1 \neq V$.

© Springer Nature Switzerland AG 2020

R. Penner, *Topology and K-Theory*, Lecture Notes in Mathematics 2262,

https://doi.org/10.1007/978-3-030-43996-5_33

Thus $Y(V) = Y^+ \cup Y^-$ where Y^+ are the layers so that $V_1 < V$ and Y^- are the layers so that $V_0 > 0$.

Remark If I is a poset and J is the poset of layers in I, then I and J are homotopy equivalent.

More generally if $\mathcal{C}$ is a small category, we can form a category $\mathcal{A}$ whose objects are arrows in $\mathcal{C}$ and in which a morphism $(X' \xrightarrow{u'} Y') \longrightarrow (X \xrightarrow{u} Y)$ is a diagram

$$
\begin{array}{ccc}
X' & \xrightarrow{u'} & Y' \\
\uparrow & & \downarrow \\
X & \xrightarrow{u} & Y
\end{array}
$$

$\mathcal{A}$ is the cofibered category over $\mathcal{C}^{op} \times \mathcal{C}$ belonging to the functor

$$
X, Y \longmapsto \mathrm{Hom}_\mathcal{C}(X, Y),
$$

and the functor $\mathcal{A} \xrightarrow{p} \mathcal{C}$ given by $(X \xrightarrow{u} Y) \longmapsto Y$ is cofibered with fiber over Y equal to $(\mathcal{C}/Y)^{op}$, which is contractible since it has an initial object.

Similarly $\mathcal{A} \xrightarrow{q} \mathcal{C}^{op}$ is cofibered with contractible fiber over $Y = X/\mathcal{C}$.

It follows that p and q are homotopy equivalences by Theorem A plus the fact that for a cofibered functor $f : \mathcal{C} \to \mathcal{C}'$ the left fiber $\mathcal{C}/Y$ is homotopy equivalent to the fiber $f^{-1}(Y)$, that is, we have

$$
f^{-1} Y \subset \mathcal{C}/Y ,
$$

$$
u_* X \longleftarrow\!\!\!\mid \left(\begin{array}{c} X \\ fX \xrightarrow{u} Y \end{array} \right) \quad \text{adjoint to inclusion.}
$$

In fact $B\mathcal{A}$ is a kind of subdivision of $B\mathcal{C}$ where $X \to Y$ subdivides to

$$
X \xrightarrow{id} X \qquad X \to Y \qquad Y \xrightarrow{id} Y
$$

and

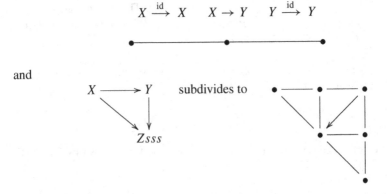

It follows that Y^+ is homotopy equivalent to the poset of all subspaces contained in but not equal to V, whence $Y^+ \cong *$. Similarly $Y^- \cong *$, and $Y^+ \cap Y^-$ is homotopy equivalent to the poset of all proper subspaces of V. Thus

$$Y(V) = \sum (Y^+ \cap Y^-)$$
$$= F_{n-1} \, Q(\mathcal{P}_A)/P \,,$$

where $\sum$ denotes the suspension.

Example The *Tits building* $T(V)$ is the simplicial complex associated to the poset of proper subspaces of V. If $G = \operatorname{Aut} V$, then the parabolic subgroups of G which are distinct from G are in one-to-one correspondence with the flags in V. We have the

Theorem (Tits) $T(V)$ *is homotopy equivalent to a bouquet of* $(n-2)$ *spheres, where n is the dimension of V.*

Example For $n = 2$, $T(V)$ is the zero-dimensional complex with vertices given by lines in V, and for $n = 3$, $T(V)$ is a connected graph.

The *Steinberg module* is $\widetilde{H}_{n-2}(T(V), \mathbb{Z})$.

Chapter 34
Bi-simplicial Sets and Dold–Thom

Theorem A $f : \mathcal{C} \to \mathcal{C}'$ *is a homotopy equivalence if* $Y \backslash f \cong *$ *for all* Y, *where* $Y \backslash f = \left\{ \left({}_{Y \xrightarrow{u} fX}^{X} \right) \right\}$ *is the cofibered category corresponding to the map* $X \mapsto \mathrm{Hom}_{\mathcal{C}'}(Y, fX)$.

Let $S(f)$ be the cofibered category over $\mathcal{C} \times (\mathcal{C}')^{\mathrm{op}}$ associated to $(X, Y) \mapsto \mathrm{Hom}_{\mathcal{C}'}(Y, fX)$, so $S(f)$ has objects $\{(X, Y, u : Y \to fX)\}$ and morphisms given by diagrams

$$
\begin{array}{ccc}
Y & \xrightarrow{\ u\ } & fX \\
\uparrow & & \downarrow \\
\big| & & \big| \\
Y' & \xrightarrow{\ u'\ } & fX'
\end{array}
$$

Let $T(f)$ be a *bi-simplicial set*, that is,

$$[p], [q] \longmapsto \text{sets of pairs}$$

$$(Y_p \longrightarrow Y_{p-1} \longrightarrow \cdots \longrightarrow Y_0 \longrightarrow fX_0, \ X_0 \longrightarrow \cdots \longrightarrow X_q)$$

of diagrams, the first in $\mathcal{C}'$ and the second in $\mathcal{C}$, and call this $T(f)_{p,q}$, i.e.,

$$
\begin{aligned}
T(f)_{p,q} &= \coprod_{X_0 \to \cdots \to X_q \in N(\mathcal{C})} N_p(\mathcal{C}'/fX_0) \\
&= \coprod_{Y_p \to \cdots \to Y_0 \in N(\mathcal{C}')} N_q(Y_0 \backslash f).
\end{aligned}
$$

© Springer Nature Switzerland AG 2020
R. Penner, *Topology and K-Theory*, Lecture Notes in Mathematics 2262,
https://doi.org/10.1007/978-3-030-43996-5_34

Proposition *Given a bi-simplicial set $p, q \mapsto T_{p,q}$, we have homeomorphisms*

$$\left| [p] \longmapsto \left| [q] \longmapsto T_{pq} \right| \right|$$

$$\approx$$

$$\left| [p] \longmapsto T_{pp} \right|$$

$$\approx$$

$$\left| [q] \longmapsto \left| [p] \longmapsto T_{pq} \right| \right|$$

Example $T_{pq} = X_p \times Y_q$ for X and Y simplicial sets, whence

$$\left| [q] \longmapsto X_p \times Y_q \right| = X_p \times |Y|$$

so

$$\left| [p] \longmapsto \left| [q] \longmapsto X_p \times Y_q \right| \right| = |X| \times |Y| \,.$$

Milnor's Theorem says $|X \times Y| = |X| \times |Y|$, where $X \times Y : [p] \longmapsto X_p \times Y_p$. Thus the Proposition works here.

Proof of Proposition First, by general nonsense, we only have to exhibit these homeomorphisms for the following generators on the category of bi-simplicial sets

$$h_{([m],[n])} : [p], [q] \longmapsto \operatorname{Hom}_{\triangle}([p], [m]) \times \operatorname{Hom}_{\triangle}([q], [n])$$

since for any $F : \mathcal{C} \to \text{Sets}$, we have

$$\varinjlim_{(X,\xi) \in \mathcal{C}_F} h_X = F \,.$$

Second, $|[p] \mapsto \operatorname{Hom}_{\triangle}([p], [m])|$ is the m-simplex $\triangle_m$, and the combinatorial part is that

$$\triangle_m \times \triangle_n = |[p] \mapsto \operatorname{Hom}_{\triangle}([p], [m]) \times \operatorname{Hom}_{\triangle}([p], [n])| \,,$$

proving the second assertion and hence the proposition. $\square$

Claim $T(f)_{pp} = N_p(S(f))$.

The proposition thus gives homeomorphisms

$$\left| p \longmapsto \left| q \longmapsto T_{pq} \right| \right| \approx \left| p \longmapsto \coprod_{Y_p \to Y_0} B(Y_0/f) \right| ,$$

$$\left| q \longmapsto \left| p \longmapsto T_{pq} \right| \right| \approx \left| q \longmapsto \coprod_{X_0 \to \cdots X_q} B(\mathcal{C}'/fX_0) \right| .$$

By the hypothesis of Theorem A

$$B(\mathcal{C}'/f X_0) \cong * \quad \text{and} \quad B(Y_0/f) \cong *.$$

We still need the

Lemma *If $X \mapsto B_X$ is a functor from $\mathcal{C}$ to topological spaces so that $B_X \cong *$, for all X in $\mathcal{C}$, then*

$$\left| [p] \longmapsto \coprod_{X_0 \to \cdots \to X_p} B_{X_0} \right| \cong \left| [p] \longmapsto \coprod_{Y_0 \to \cdots \to X_p} \text{point} \right| = B\mathcal{C}.$$

To see this, consider $\left| [p] \mapsto \coprod\limits_{X_0 \to \cdots \to X_p} B_{X_0} \right|$ and conclude that

$$\left| [p] \longrightarrow \coprod_{Y_p \to \cdots Y_0} B(Y_0/f) \right| \xrightarrow{\simeq} B\mathcal{C}',$$

$$\left| [q] \longrightarrow \coprod_{X_0 \to \cdots \to X_q} B(\mathcal{C}'/f X_0) \right| \xrightarrow{\simeq} B\mathcal{C}.$$

Thus we find

$$B(S(f)) \xrightarrow{\simeq} B\mathcal{C}'$$
$$\downarrow \simeq$$
$$B\mathcal{C}$$

and compare this with the same thing for $\mathrm{id}_{\mathcal{C}'} : \mathcal{C}' \to \mathcal{C}'$:

$$\mathcal{C}' \longleftarrow S(f) \longrightarrow \mathcal{C}$$
$$\| \qquad \qquad \qquad \downarrow f$$
$$\mathcal{C}' \longleftarrow S(\mathrm{id}_{\mathcal{C}}') \longrightarrow \mathcal{C}'$$

where the right-hand square of the diagram commutes. The argument above is natural in f thus proving Theorem A. □

Recall that if $E \xrightarrow{f} B$ is a map of spaces and $b \in B$, then the *homotopy fiber* of f over b is

$$E(f; b) = E \times_B B^I \times_B \{b\}$$
$$= \{(e, \lambda) : \lambda \text{ is a path from } f(e) \text{ to } b\}.$$

This is the actual fiber over b of the Serre construction

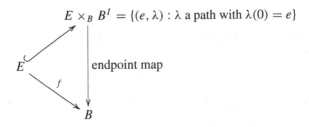

$$E \times_B B^I = \{(e, \lambda) : \lambda \text{ a path with } \lambda(0) = e\}$$

endpoint map

We get of course $e \in f^{-1}(b)$ and the homotopy exact sequence

$$\cdots \longrightarrow \pi_{n+1} \longrightarrow \pi_n(E(f; b), \widetilde{e}) \longrightarrow \pi_n(E, e) \longrightarrow \pi_n(B, b) \longrightarrow \pi_{n-1} \longrightarrow \cdots.$$

Note that

$$f^{-1}b \subset E(f, b)$$

via

$$\widetilde{e} = (e, \text{constant path})$$

We call f a *quasi-fibration* (in the sense of Dold–Thom) when the inclusion $f^{-1}b \subset E(f, b)$ is a weak homotopy equivalence for all $b \in B$.

Dold–Thom Example Let A be a "nice" subspace of a "nice" space X, where A is connected and has a base point, and let $SP(X) = \bigcup_n X^n / \Sigma^n$ be the infinite symmetric product. We have also $SP(X/A)$, and the map

$$SP(X) \longrightarrow SP(X/A)$$

is a quasi-fibration (this is a theorem), where the fiber over the base point is $SP(A)$. We conclude that there is an exact sequence

$$\cdots \longrightarrow \pi_n SPA \longrightarrow \pi_n SPX \longrightarrow \pi_n SPX/A \longrightarrow \cdots$$

We also have the

Theorem $\pi_n SPX = \widetilde{H}_n(X, \mathbb{Z})$.

A good paradigm for a quasi-fibration which is not a fibration is given by the natural mapping from the mapping cylinder of $X \xrightarrow{f} Y$ to the unit interval, which is not a fibration unless f is the identity map but is a quasi-fibration provided f is a homotopy equivalence.

Theorem B *Suppose $f : C \to C'$ is so that for all $Y' \xrightarrow{u} Y$ in C', the functor $u^* : Y \backslash f \longrightarrow Y' \backslash f$ is a homotopy equivalence. Then $B(Y \backslash f)$ is homotopy equivalent to the homotopy fiber of Bf over Y.*

Proof We have

$$\left| [p] \longmapsto \coprod_{Y_p \to \cdots \to Y_0} B(Y_0 \backslash f) \right|$$

$$\downarrow$$

$$\left| [p] \longmapsto \coprod_{Y_p \to \cdots \to Y_0} \text{point} \right|$$

This is a quasi-fibration from Dold–Thom theory because all specialization maps are homotopy equivalences by assumption.

As in Theorem A, we have

$$\left| [p] \longrightarrow \coprod_{Y_p \to \cdots \to Y_0} B(Y_0 \backslash f) \right| = \left| [q] \longmapsto B(C'/X_0) \right| \simeq BC$$

and

$$\left| [p] \longmapsto \coprod_{Y_p \to \cdots \to Y_0} \text{point} \right| \simeq BC'. \qquad \square$$

Chapter 35
Homology of $Q(\mathcal{P}_A)$ and the Tits Complex

We begin with the

Example Take

$$SPX \xrightarrow{f} SPX/A$$

and let $[y_1, \cdots, y_n] \in SP(X/A)$, where without loss of generality $y_i \in X \backslash A$. Then $f^{-1}[y_1, \cdots, y_n]$ is the union of $[y_1, \cdots, y_n]$ and any set in $SP(A)$. Specialization maps are multiplication by elements of $SP(A)$. $SP(A)$ is a connected monoid, so these are homotopy equivalences.

Let A be a Dedekind domain and F its field of fractions. We defined $F_n Q(\mathcal{P}_A)$ to be the full subcategory comprised of those P with rank at most n. Then

$$0 = F_0 \subset F_1 \subset \cdots \subset \bigcup_n F_n = Q(\mathcal{P}_A).$$

Problem Compare the homology of F_n and F_{n-1}. To simplify what follows, assume $\text{Pic}(A) = 0$, so the only projective module P of rank n is A^n and

$$\text{Hom}_{Q(\mathcal{P}_A)}(P, P) = \text{Aut}(P) = GL_n(A).$$

If we regard $\text{Aut} A^n$ as a category, then we have inclusions

$$F_{n-1} \xhookrightarrow{i} F_n \xleftarrow{\quad}_{j} \text{Aut} A^n$$

of categories. Recall from last term that for a functor $f : \mathcal{C} \to \mathcal{C}'$ and a functor $F : \mathcal{C} \to \text{Ab}$, we have

$$f_!(F)(Y) = \operatorname*{colim}_{(X, fX \to Y)} F(X).$$

© Springer Nature Switzerland AG 2020
R. Penner, *Topology and K-Theory*, Lecture Notes in Mathematics 2262,
https://doi.org/10.1007/978-3-030-43996-5_35

More generally, if F. is a complex in Funct $(\mathcal{C}, \mathrm{Ab})$, then we have a complex $\mathbb{L}f_!(F.)$ in Funct $(\mathcal{C}', \mathrm{Ab})$ unique up to quasi-isomorphism. There is an isomorphism

$$H_p(\mathcal{C}', \mathbb{L}f_!(F.)) \cong H_p(\mathcal{C}, F.),$$

and there is a spectral sequence with

$$E_{pq}^2 = H_p(\mathcal{C}', L_q f_!(F.)) \text{ converging to } H_n(\mathcal{C}', \mathbb{L}f_!(F.))$$

with

$$L_q f_!(P)(Y) = H_q(f/Y, F \text{ pulled back to } f/Y).$$

Setting $\Gamma = \mathrm{GL}_n A$, if we put $f = j : \widetilde{\Gamma} \hookrightarrow F_n$, then

$$j/p = \begin{cases} \varnothing, \text{ if } P \in F_{n-1}, \\ \text{the category whose objects are elements of } \mathrm{Hom}(A^n, P) \text{ and whose} \\ \text{maps are given by composition with elements of } \Gamma, \text{ if rank } P = n. \end{cases}$$

So if M is any Γ-module, then

$$L_q j_!(M)(P) = \begin{cases} 0, & \text{if } P \in F_{n-1}, \\ 0, & \text{if } P \cong A^n \text{ and } q > 0, \\ M, & \text{if } P \cong A^n \text{ and } q = 0, \end{cases}$$

so

$$H_*(F_n, j_!(M)) = H_*(\Gamma, M).$$

Note that $j_!(M)$ is a typical functor on F_n with support $\{A^n\}$.

Now we consider $i : F_{n-1} \to F_n$. Then for $P \in F_{n-1}$, i/p is a category with final object, hence $H_q(i/p, F) = 0$, for $q > 0$, and any $F : i/p \to \mathrm{Ab}$. i/A^n is equivalent to the poset of proper layers in F^n, which in turn is equivalent to the suspension of $T(F^n)$, namely the poset of proper subspaces of F^n. Thus

$$\widetilde{H}_k(i/A^n, \mathbb{Z}) = \widetilde{H}_{k-1}(T(F^n), \mathbb{Z})$$
$$= \begin{cases} 0, & k-1 \neq n-2, \\ I(F^n), & k-1 = n-2, \end{cases}$$

where $I(F^n)$ is the Steinberg module of F^n. Now we have

$$H_*(F_{n-1}, \mathbb{Z}) = H_*(F_n, \mathbb{L}i_!(\mathbb{Z})),$$

$$L_q i_!(\mathbb{Z})(p) = \begin{cases} 0, q \neq 0, \ p \in F_{n-1}, \\ \mathbb{Z}, q = 0, \ p \in F_{n-1}, \end{cases}$$

and

$$L_q \, i_!(\mathbb{Z})(A^n) = H_q(i/A^n, \mathbb{Z}) = \begin{cases} 0, & q \neq 0, n-1, \\ \mathbb{Z}, & q = 0, \\ I(F^n), & q = n-1, \end{cases}$$

provided $n \neq 1$. A formula valid even for $n = 1$ is an exact sequence

$$0 \longleftarrow \mathbb{Z} \longleftarrow \mathbb{L}\, i_!(\mathbb{Z}) \longleftarrow j_! I(F^n)[n-1] \longleftarrow 0 \,,$$

and from this we get exact homology sequences

$$\cdots \longleftarrow H_q(F_n, \mathbb{Z}) \longleftarrow H_q(F_n, \mathbb{L}\, i_!(\mathbb{Z})) \longleftarrow H_q(F_n, j_! \, I(F^n)[n-1]) \longleftarrow \cdots$$

$$\cdots \longleftarrow H_q(F_n, \mathbb{Z}) \longleftarrow H_q(F_{n-1}, \mathbb{Z}) \longleftarrow H_{q-n+1}(GL_n(A), I(F^n)) \longleftarrow \cdots$$

The Tits complex

For a vector space V, $|T(V)|$ is the simplicial complex associated to the poset $T(V)$ of proper subspaces of V. Choose a line L in V and consider the map $W \mapsto W + L$. If this were a well-defined map $T(V) \to T(V)$, then $T(V)$ would be conically contractible since we would have

$$W \leq W + L \geq L \,.$$

But the map fails to be well-defined on

$$\mathcal{H}_L = \{\text{hyperplanes } H \text{ with } H + L = V\} \,,$$

and the above argument shows that the poset $T(V) - \mathcal{H}_L$ is contractible.

Now for a vertex v in a simplicial complex K, we define

Link v = the subcomplex of simplices τ with $v \notin \tau$ such that $v \cup \tau \in K$,

Star v = the subcomplex of simplices σ such that $v \in \sigma$,

$\overline{\text{Star}}\, v$ = the subcomplex of simplices σ with $v \cup \tau \in K$,

so we have

$\overline{\text{Star}}\, v = \text{Star } v \cup \text{Link } v$.

Then for any v, we have

$$K = (K - \text{Star } v) \bigcup_{\text{Link } v} \overline{\text{Star}}\, v,$$

so

$$|T(v)| = \bigcup_{H \in \mathcal{H}_L} |T(v) - \mathcal{H}_L| \cup \bigcup_{\text{Link } H} \text{cone on Link } (H)$$

since no two H in $\mathcal{H}_L$ form a simplex. Then since $|T(v) - \mathcal{H}_L| \cong *$, we find

$$|T(v)| \simeq \bigcup_{H \in \mathcal{H}_L} S \operatorname{Link}(H).$$

But Link $(H) = |T(H)|$, so by induction $|T(v)|$ is a bouquet of spheres, and

$$\#\text{spheres in } |T(P^n)| = (\#H \text{ in } \mathcal{H}_L) \cdot (\#\text{spheres in } |T(P^{n-1})|).$$

Moreover, for F a finite field of characteristic q, we have $\#H$ in $\mathcal{H}_L = q^n$.

It follows from this argument that once a flag $0 < F_1 < F_2 < \cdots < F_{n-1} < V$ is chosen, then $I(V)$ has a basis indexed in a natural way by flags

$$V > H_1 > H_2 > \cdots > H_{n-1} > 0$$

with $H_i \oplus F_i = V$ for each i. As a module over the unipotent radical of the Borel subgroup B^n of Aut $(V) = GL_n(F)$ fixing $\{F_i\}$, $I(V)$ is free, i.e.,

$$I(V) \simeq \mathbb{Z}[B^n].$$

For $F = \mathbb{F}_q$, B^n is a Sylow p-subgroup and $I(V)$ is projective over Sylow subgroups. It follows that $I(K)_p$ is projective over $\mathbb{Z}_p[GL_n(F)]$.

Chapter 36
Long Exact Sequences of K-Groups

Recall

Theorem A: $f : \mathcal{C} \to \mathcal{C}'$ is a homotopy equivalence provided any of the following hold for all Y in $\mathcal{C}'$

(i) $f/Y \simeq *$,
(i') f is pre-cofibered and $f^{-1}Y \simeq *$,
(ii) $Y \backslash f \simeq *$,
(ii') f is pre-fibered and $f^{-1}Y \simeq *$.

Pre-cofibered means "there are cobase change operations," i.e., given $u : Y \to Y'$ in $\mathcal{C}'$, we have a *cobase change functor* $u_* : f^{-1}(Y) \to f^{-1}(Y')$, that is, $f^{-1}Y \hookrightarrow f/Y$ has an adjoint, namely, $(X, fX \overset{u}{\longmapsto} Y) \longmapsto u_*(X))$.

Suppose $\mathcal{M}$ is an exact category. Let $\mathcal{E}(\mathcal{M})$ denote the category of short exact sequences $M' \hookrightarrow M \twoheadrightarrow M''$ in $\mathcal{M}$. $\mathcal{E}(\mathcal{M})$ is itself an exact category, and we have three exact functors s, t, q from $\mathcal{E}(\mathcal{M})$ to $\mathcal{M}$:

$$s(M' \hookrightarrow M \twoheadrightarrow M'') = \text{sub-object} = M',$$
$$t(M' \hookrightarrow M \twoheadrightarrow M'') = \text{total object} = M,$$
$$q(M' \hookrightarrow M \twoheadrightarrow M'') = \text{quotient object} = M''.$$

Remark Exact functors induce maps on Q-categories and hence on K-groups.

Theorem $(s, q) : \mathcal{E}(\mathcal{M}) \to \mathcal{M} \times \mathcal{M}$ *induces a homotopy equivalence on Q-categories, i.e.,* $Q(\mathcal{E}(\mathcal{M})) \simeq Q\mathcal{M} \times Q\mathcal{M}$.

(†) **Corollary:** Given an exact category $\mathcal{N}$ and an exact sequence of exact functors $0 \to F' \to F \to F'' \to 0$ from $\mathcal{N}$ into $\mathcal{M}$. Then on K-groups, we have $F_* = F'_* + F''_* : K_i(\mathcal{N}) \to K_i(\mathcal{M})$.

Proof We have the diagram

© Springer Nature Switzerland AG 2020

R. Penner, *Topology and K-Theory*, Lecture Notes in Mathematics 2262,
https://doi.org/10.1007/978-3-030-43996-5_36

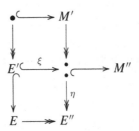

and a section η of $\mathcal{E}(\mathcal{M}) \xrightarrow{(s,q)} \mathcal{M} \times \mathcal{M}$ by

$$(M', M'') \mapsto (M' \hookrightarrow M' \oplus M'' \twoheadrightarrow M'').$$

Thus adding $Q(\eta)$ to the diagram, we get the result. $\square$

Proof of Theorem It is sufficient by Theorem A to show that for any $(M', M'') \in QM \times QM$, we have $Q(\mathcal{E}(M))(M', M'') \simeq *$. The typical element is

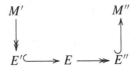

By pulling back with respect to η and pushing forward by ξ, we retract $Q(\mathcal{E}(M))/$ (M', M'') into the full subcategory of diagrams

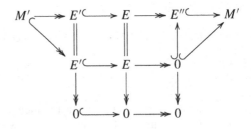

But this full subcategory has the initial object

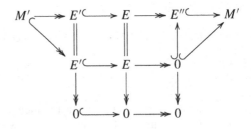

                                                                        $\square$

Theorem B: Let $f : C \to C'$ and suppose that for any $Y \to Y'$ in C', the induced functor $f/Y \to f/Y$ is a homotopy equivalence. Then the homotopy fiber of $Bf : BC \to BC'$ over Y is homotopy equivalent to $B(f/Y)$.

Geometrical Example Let $g : K \to L$ be a map of simplicial complexes. Take C to be the poset of simplices in K, C' to be the poset of simplices in L and suppose y is a simplex in L. Then

$$
\begin{aligned}
f/y &= \{\text{simplices } x \text{ in } K \;:\; fx \subset y\} \\
&= \{\text{simplices in } f^{-1}\,\overline{y}\},
\end{aligned}
$$

and $f^{-1}\,y' \simeq f^{-1}\,y$ for all $y' \le y$ implies that $f^{-1}\,y \simeq$ homotopy fiber.

Localization Theorem *Suppose $\mathcal{A}$ is an abelian category and $\mathcal{B}$ a Serre subcategory with $\mathcal{A}/\mathcal{B}$ the quotient abelian category. Then*

$$
BQ\mathcal{B} \simeq \text{homotopy fiber of } (BQ\mathcal{A} \longrightarrow BQ(\mathcal{A}/\mathcal{B})).
$$

Proof From Theorem B plus lots of work, we have a short exact sequence

$$
Q\mathcal{B} \longrightarrow Q\mathcal{A} \longrightarrow Q(\mathcal{A}/\mathcal{B}).
$$

See Quillen's paper. $\square$

Corollary *There is an exact sequence of K-groups*

$$
\cdots \longrightarrow K_i\,\mathcal{B} \longrightarrow K_i\,\mathcal{A} \longrightarrow K_i\,\mathcal{A}/\mathcal{B} \longrightarrow K_{i-1}\,\mathcal{B} \longrightarrow \cdots
$$

$\square$

Remark Suppose $V \in \mathcal{A}/\mathcal{B}$. We have

$$
SB\,\mathrm{Aut}\,V \longrightarrow BQ(\mathcal{A}/\mathcal{B}),
$$

so

$$
B\,\mathrm{Aut}\,V \longrightarrow \Omega\,BQ(\mathcal{A}/\mathcal{B}),
$$

and we get

$$
\Omega BQ(\mathcal{A}/\mathcal{B}) \longrightarrow BQ\mathcal{B} \longrightarrow BQ\mathcal{A} \longrightarrow BQ(\mathcal{A}/\mathcal{B}).
$$

$B\,\mathrm{Aut}\,V$

We exhibit this map. Given $V \in \mathcal{A}/\mathcal{B}$, take $\mathcal{A} = \mathrm{Modf}\,A$, $\mathcal{B} = S$-torsion and $\mathcal{A}/\mathcal{B} = \mathrm{Modf}(S^{-1}A)$, where A is a noetherian ring.

Construction: Take the poset JV of all finitely generated A-submodules in $M \subset V$ so that $V = S^{-1}(M)$. $J(V) \simeq *$ because we can take suprema. Let $\tilde{J}V =$ poset of layers in JV. We have a functor

$$\tilde{J}V \longrightarrow Q\mathcal{B}$$
$$(M_0, M_1) \longmapsto M_1/M_0\,.$$

Now form the cofibered category $\tilde{J}V$ Aut V over Aut V with fiber $\tilde{J}V$. The objects are still (M_0, M_1) but now morphisms

$$(M_0', M_1') \longrightarrow (M_0, M_1)$$

are automorphisms θ of V so that $(\theta M_0', \theta M_1') \leq (M_0, M_1)$. We have a functor

$$\tilde{J}V\,\text{Aut}\,V \xrightarrow{\hspace{3cm}} Q\mathcal{B}$$

$$\downarrow {\scriptstyle \text{homotopy equivalence by Theorem A}}$$

$$\text{Aut}\,V$$

Example 1 Recall if A is regular noetherian, then

$$K_0\,A \cong K_0\,A[T]\,.$$

This generalizes with a proof as before, so

$$K_n\,A \cong K_n\,A[T]\,.$$

Example 2 For A regular, consider $K_*\,A[T, T^{-1}]$. We have

$$K_*(\mathcal{P}_A) = K_*(\text{Modf}\,A) \text{ for } A \text{ and also for } A[T]$$

and consider

$$\text{Modf}\,A[T] \longrightarrow \text{Modf}\,A([T]) \longrightarrow \text{Modf}\,A[T, T^{-1}]\,.$$
$$\begin{array}{ccc} T\text{-torsion} & & \\ \mathcal{B} & \mathcal{A} & \mathcal{A}/\mathcal{B} \end{array}$$

By Devissage, $\mathcal{B}$ has the same K-groups as $A[T]$-modules killed by T, and so

$$K_*\,\mathcal{B} = K_*\,A\,.$$

We get the long exact sequence

$$\cdots \longrightarrow K_i\,A \xrightarrow{\ \alpha\ } K_i\,A[T] \longrightarrow K_i\,A[T, T^{-1}] \longrightarrow K_{i-1}\,A \longrightarrow \cdots .$$

$$\Big\Vert$$

$$K_i\,A$$

To compute α, take $M \in \mathrm{Modf}\,A \subset \mathcal{B}$. We have the exact sequence

$$0 \longrightarrow \overbrace{A[T]\otimes_A M}^{F} \xrightarrow{\ \times T\ } A[T] \otimes_A M \longrightarrow M \longrightarrow 0$$

of exact functors from $\mathrm{Modf}\,A$ to $\mathrm{Modf}\,A[T]$.

By Corollary (†), we conclude that $\alpha = 0$, i.e., $F_* = F_* + \alpha$, and so finally

$$K_i\,A[T, T^{-1}] \cong K_i\,A \oplus K_{i-1}\,A\,.$$

Chapter 37
Localization

Let

$$X = \text{algebraic variety over a field } k,$$
$$\mathcal{M}_X = \text{(abelian) category of coherent sheaves on } X,$$
$$\mathcal{P}_X = \text{category of locally free sheaves on } X.$$
$$(\text{“vector-bundles”})$$

If

$$X = \text{Spec } A, \text{ for } A \text{ a finitely generated } k\text{-algebra},$$
$$\mathcal{M}_X = \text{finitely generated modules over } A,$$
$$\mathcal{P}_X = \text{finitely generated projective modules over } A,$$

then $K_i\, X \overset{d}{=} K_i\, \mathcal{P}_X$ is a contravariant functor in X because $f : X \to Y$ induces an exact functor $f^* : \mathcal{P}_Y \to \mathcal{P}_X$.

Let us also define $K_i'\, X \overset{d}{=} K_i\, \mathcal{M}_X$, a covariant functor in X for proper maps. This holds because of Grothendieck's argument for $i = 0$: If $f : X \to Y$ and F is a coherent sheaf on X, then $R^q\, f_*\, F$ are coherent sheaves on Y and vanish for $q > \dim X$. So we can form

$$\sum_{q \geq 0} (-1)^q\, [R^q\, f_*\, F] \in K_0\, \mathcal{M}_Y.$$

This gives a map $\mathcal{M}_X \to K_0\, \mathcal{M}_Y$ which is additive for short exact sequences, namely, the Euler characteristic, and so induces $K_0\, \mathcal{M}_X \to K_0\, \mathcal{M}_Y$.

For higher K-groups, assume that X is *quasi-projective*, i.e., X can be embedded in projective space, so a proper morphism $f : X \to Y$ is projective, i.e., factors as a composition $X \overset{\epsilon}{\hookrightarrow} Y \times P^n \overset{p}{\to} Y$ with $i\epsilon$ a closed embedding and p the projection.

© Springer Nature Switzerland AG 2020

R. Penner, *Topology and K-Theory*, Lecture Notes in Mathematics 2262,
https://doi.org/10.1007/978-3-030-43996-5_37

Then the map $\epsilon_* : K_i'(X) \to K_i'(Y \times P^n)$ is induced by the embedding $\mathcal{M}_X \to \mathcal{M}_{Y \times P^n}$. The map $p_* : K_i(Y \times P^n) \to K_i(Y)$ can be defined using Serre's theorem

$$R^q \, p_*(F(n)) = 0 \, , \text{ for } q > 0$$

where

$$F(n) = F \otimes \mathcal{O}(1)^{\otimes n} \, .$$

This shows that large chunks of $\mathcal{M}_{Y \times P^n}$ (and thus of $\mathcal{M}_X$) map, by $R^0 p_*$ (resp. by $R^0 f_*$) into $\mathcal{M}_Y$, in a way preserving short exact sequences, and that is sufficient. For a more detailed and sophisticated explanation see §6.2 of Quillen's paper [10].

Suppose $A \xrightarrow{f} B$ is a map between noetherian rings. This induces an exact functor $\mathcal{P}_A \to \mathcal{P}_B$, where $P \mapsto P \otimes_A B$, and hence a map $K_i \, A \to K_i \, B$.

However in general, the map Modf $A \to$ Modf B given by $M \mapsto B \otimes_A M$ is not exact. But assume that B_A has *finite Tor dimension*, i.e., the ith left derived functor $\mathrm{Tor}_i(B, M) = 0$ for all sufficiently large i. Then we can define a map

$$K_0 \, \mathrm{Modf} \, A \longrightarrow K_0 \, \mathrm{Modf} \, B$$
$$M \longmapsto \sum_{q \geq 0} (-1)^q [\mathrm{Tor}_i^A(B, M)] \, .$$

For higher K-groups, introduce a filtration

$$F_0 \, \mathcal{M}_A \subset \cdots \subset F_p \, \mathcal{M}_A \subset \cdots \subset \mathrm{Modf} \, A = \mathcal{M}_A \, ,$$

where $M \in F_p \, \mathcal{M}_A$ if and only if $\mathrm{Tor}_i^A(B, M) = 0$ for $i > p$. Then

$$M \mapsto \mathrm{Tor}_0 = B \otimes_A M$$

is exact from $F_0 \, \mathcal{M}_A$ to $\mathcal{M}_B$ and so induces a map

$$K_i \, F_0 \, \mathcal{M}_A \longrightarrow K_0 \, \mathcal{M}_B \, .$$

Now, we use the

Resolution Theorem *Suppose we have a category closed under extensions*

$$0 \longrightarrow M' \longrightarrow M \longrightarrow M'' \longrightarrow 0 \, .$$

Then we have an induced exact sequence

$$\cdots \longrightarrow \mathrm{Tor}_p^A(B, M') \longrightarrow \mathrm{Tor}_p^A(B, M) \longrightarrow \mathrm{Tor}_p^A(B, M'') \longrightarrow \mathrm{Tor}_{p-1}^A(B, M') \longrightarrow \cdots \, .$$

Moreover, if M in a subcategory implies that M' is also in the subcategory, then we find

$$K_i \, F_{p-1} \, \mathcal{M}_A \cong K_i \, F_p \, \mathcal{M}_A$$

and get an induced map in higher K-groups.

Example of Localization Theorem Let $\mathbb{Q}$ be the quotient field of $\mathbb{Z}$. Then we have the exact sequence

$$\mathcal{T} = \text{finitely generated torsion } \mathbb{Z}\text{-modules} \longrightarrow \mathcal{M}_{\mathbb{Z}} \longrightarrow \mathcal{M}_{\mathbb{Q}} \, ,$$

and

$$K_* \, \mathcal{M}_{\mathbb{Z}} = K_* \, \mathbb{Z} \text{ since } \mathbb{Z} \text{ is regular by the Resolution Theorem}$$

and similarly for $K_* \, \mathbb{Q}$.
Now

$$\mathcal{T} = \prod_p \mathcal{T}_p \, ,$$

where the product is over primes p and $\mathcal{T}_p$ denotes the p-torsion finite abelian groups, whence

$$K_* \, \mathcal{T} = \bigoplus_p K_* \, \mathcal{T}_p \, .$$

Use Devissage to get

$$K_* \, \text{Modf}(\mathbb{Z}/p\,\mathbb{Z}) \approx K_* \, \mathcal{T}_p \, ,$$

whence from localization

$$\cdots \longrightarrow \bigoplus_p K_i \, \mathbb{Z}/p \longrightarrow K_i \, \mathbb{Z} \longrightarrow K_i \, \mathbb{Q} \longrightarrow \bigoplus_p K_{i-1} \, \mathbb{Z}/p \longrightarrow \cdots \, .$$

There is an obvious generalization

$$\cdots \longrightarrow \bigoplus_m K_i \, A/m \longrightarrow K_i \, A \longrightarrow K_i \, F \longrightarrow \bigoplus_m K_{i-1} A/m \longrightarrow \cdots$$

to a Dedekind domain A with quotient field F.

Facts:

$$[\text{Quillen}] \quad K_i(\mathbb{F}_q) = \begin{cases} \mathbb{Z} \, , & i = 0 \, , \\ \mathbb{F}_q^* \cong \mathbb{Z}/(q-1) \, , & i = 1 \, , \\ 0 \, , & i \text{ even} \, , \\ \mathbb{Z}/(q^j - 1) \, , & i = 2j - 1 \text{, for } j \geq 1 \, . \end{cases}$$

Suppose A is the ring of integers in a number field F with r_1 real places and r_2 complex places. Then

$$[\text{Borel}] \quad \text{rank } K_i A = \begin{cases} 1, & i = 0, \\ r_1 + r_2 - 1, & i = 1, \\ 0, & i = \text{even}, \\ r_2, & i = 3, 7, 11, 15, \ldots, \\ r_1 + r_2, & i = 5, 9, 13, \ldots. \end{cases}$$

A general fact [Bass] is that $K_1 A = A^*$ for integers in number fields.

From [Quillen], we get

$$K_i(\overline{\mathbb{F}}_p) = \begin{cases} \mathbb{Z}, & i = 0, \\ 0, & i \equiv 0(2), \\ \bigoplus\limits_{\substack{\ell \neq p \\ \text{prime}}} \mathbb{Q}_\ell/\mathbb{Z}_\ell, & i \text{ odd}, \end{cases}$$

and in fact $\mathbb{Q}_\ell/\mathbb{Z}_\ell = \bigcup_{n \geq 0} \mathbb{Z}/\ell^n \mathbb{Z}$.

Example Suppose $k = \overline{\mathbb{F}}_p$ and let X be a complete non singular curve over k. We have

$$\begin{array}{ccc} k[T^{-1}] & \subset & A_- \\ \cap & & \cap \\ k(T) & \subset & F \\ \cup & & \cup \\ k[T] & \subset & A_+ \end{array}$$

where $A_\pm$ is the integral closure of $k[T^{\pm 1}]$ in F, and we set $B = A_+ A_- \subset F$.

Then $X = \text{Spec } A_+ \cup \text{Spec } A_-$ is a finite ramified cover of $\mathbb{P}_1(k)$, and we have $\mathcal{P}_X = (P_+, P_-, \alpha)$, for $P_\pm \in \mathcal{P}_{A_\pm}$, where α an isomorphism of $P_+ \otimes B$ with $P_- \otimes B$.

This is like a Dedekind domain where we have the following "localization"

$$\cdots \longrightarrow \bigoplus_{\substack{\text{closed points} \\ x \text{ in } X}} K_i\, k(x) \longrightarrow K_i\, X \longrightarrow K_i\, F \longrightarrow \cdots,$$

and note that $k(x) \cong \overline{\mathbb{F}}_p$ because k is algebraically closed.

Take

$$D = \text{the group of divisors on curve}$$
$$= \text{free abelian group on closed points},$$

whence

$$\bigoplus_{x \in X} K_i(k) = D \bigotimes_{\mathbb{Z}} K_i(k).$$

This gives

$$\cdots \longrightarrow D \otimes K_i\, k \longrightarrow K_i\, X \longrightarrow K_i\, F \longrightarrow D \otimes K_{i-1}\, k \longrightarrow \cdots ,$$

ending

$$K_1\, F \longrightarrow D \otimes K_0\, k \longrightarrow K_0\, X \longrightarrow K_0\, F \longrightarrow 0$$

$$F^* \longrightarrow D$$

with this bottom map associating to F its divisor

$$D = \mathbb{Z} \bigoplus D_0$$

of zeros and poles, where D_0 are divisors with total degree zero and the summand $\mathbb{Z}$ captures the degree. Let J be the Jacobian of the curve. Then it is well-known that $D_0/\operatorname{Im} F^* = J$, and so

$$K_0\, X = \underset{\text{rank}}{\mathbb{Z}} \bigoplus \underset{\text{degree}}{\mathbb{Z}} \bigoplus J .$$

Note that J is divisible.

Chapter 38
The Plus Construction, K_1 and K_2

An *acyclic map* $f : X \to Y$ is one so that either of the following two equivalent conditions holds

(i) the homotopy fiber of f is acyclic,
(ii) for all local coefficient systems L, we have $H_*(X, f^{-1} L) \simeq H_*(Y, L)$.

Theorem *Given X and a perfect normal subgroup N of $\pi_1 X$, there is a unique acyclic map $f : X \to Y$ with kernel N. Also, there is the following universal property in the homotopy category*

with a unique h making the diagram commute if and only if $N < \operatorname{Ker} \pi_1 g$.

Recall that the set of acyclic maps is closed under push out and pull back.

Take $X = BGL(A)$ and so $\pi_1 X = GL(A)$ with $\pi_q X = 0$ for all $q \geq 2$. Take $N = E(A)$, which is perfect and the commutator subgroup by Whitehead. Denote the unique acyclic map $f : X \to Y$ with $\operatorname{Ker} \pi_1 f = E(A)$ by

$$f : BGL(A) \longrightarrow BGL(A)^+ .$$

Review of $K_1 A$ [Bass] and $K_2 A$ [Milnor], see Milnor's book:

According to Bass, K_1 is

$$K_1 A \overset{d}{=} GL(A)^{ab} = GL(A)/E(A) = \pi_1(BGLA^+).$$

© Springer Nature Switzerland AG 2020
R. Penner, *Topology and K-Theory*, Lecture Notes in Mathematics 2262,
https://doi.org/10.1007/978-3-030-43996-5_38

One way of defining Milnor's K_2 is

$$K_2 A \overset{d}{=} H_2(E(A), \mathbb{Z})$$
$$= H_2(BEA, \mathbb{Z}) .$$

Meanwhile we have the pull back

$$
\begin{array}{ccc}
BEA = \overline{BGLA} & \xrightarrow{\text{acyclic}} & \widetilde{BGLA^+} \\
\downarrow & & \downarrow {\scriptstyle K_1} \\
BGLA & \xrightarrow{\hspace{3cm}} & BGLA^+
\end{array}
$$

and

$$\pi_2 BGLA^+ = \pi_2 \widetilde{BGLA^+}$$
$$= H_2 \widetilde{BGLA^+} , \text{ by Hurewicz}$$
$$= H_2 \overline{BGLA}$$
$$= H_2 BEA .$$

Note that $\Omega B Q \, \mathcal{P}_A = K_0 A \times BGLA^+$.

The Schur multiplier and the universal central extension of a perfect group

Recall from day one that $H^2(G, M)$ is the collection of extensions

$$* \longrightarrow M \longrightarrow E \longrightarrow G \longrightarrow *$$

of G by M, and the extension is central if and only if G acts trivially on M. If M is a trivial G-module, then we have a Universal Coefficient Theorem

$$0 \longrightarrow \operatorname{Ext}^1(H_1 G, M) \longrightarrow H^2(G, M) \longrightarrow \operatorname{Hom}(H_2 G, M) \longrightarrow 0 ,$$

and G perfect implies $\operatorname{Ext} = 0$, so

$$H^2(G, M) = \operatorname{Hom}(H_2 G, M) .$$

Translation: there exists a canonical central extension $\xi \in H^2(G, H_2 G)$ so that any other extension is induced by a unique map from $H_2 G$ to M:

$$
\begin{array}{ccccccccc}
\xi : & * & \longrightarrow & H_2 G & \longrightarrow & \widetilde{G} & \longrightarrow & G & \longrightarrow & * \\
& & & \downarrow & & \downarrow & & \| & & \\
& * & \longrightarrow & M & \longrightarrow & E & \longrightarrow & G & \longrightarrow & *
\end{array}
$$

For G perfect,

$\widetilde{G}$ is the *universal central extension of G,* and

$H_2\, G$ is the *Schur multiplier of G,*

and in particular for $E\, A$,

$\widetilde{E\, A}$ is the *Steinberg group,* and

$$H_2\, E\, A = \mathrm{Ker}\, \{\widetilde{E\, A} \longrightarrow E\, A\} = K_2\, A.$$

$E\, A$ is generated by $I + a\, E_{ij} = e_{ij}(a)$, and the commutators are

$$[e_{ij}\, a,\, e_{jk}\, b] = e_{ik}(ab)\, , \ \text{ for } i,\, j,\, k \text{ pairwise distinct.}$$

In fact, Milnor uses this to show $K_2(\mathbb{F}_q) = 0$.

Exercise $\pi_3(BGL\, A^+) = H_3\, (\widetilde{E\, A},\, \mathbb{Z})$.

$BGLA$ is not an H-space since π_1 is not abelian. Put $\mathbb{N} = \{1,\, 2,\, \ldots\}$ and let $\theta : \mathbb{N} \hookrightarrow \mathbb{N}$ be any embedding. Then θ induces a map $GLA \to GLA$ given by

$$\alpha_{ij} \mapsto \begin{cases} \delta_{ij}\, , & \text{if } i \text{ or } j \notin \mathrm{Im}\, \theta\, , \\ \alpha_{\theta^{-1}(i)\theta^{-1}(j)}\, , & \text{else,} \end{cases}$$

which for instance simply shifts α diagonally $\alpha \mapsto \begin{pmatrix} 1 & 0 \\ 0 & \alpha \end{pmatrix}$ in case $\theta(n) = n + 1$. We get

$$\begin{array}{ccc} BGLA & \longrightarrow & BGLA \\ \downarrow & & \downarrow \\ BGLA^+ & \underset{\theta}{-\, -\, \to} & BGLA^+ \end{array}$$

and have

Lemma $\theta \simeq \mathrm{id}$.

Proof
Step 1: Show the homotopy equivalence. We have

$$\begin{array}{ccc} GL_n\, A & \longrightarrow & GLA \\ & & \downarrow{\scriptstyle \theta} \\ & & GLA \end{array}$$

and there exists an $\alpha \in E\, A$ so that conjugation by α and the map θ have the same effect on $GL_n\, A$. Thus the diagram

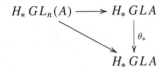

commutes because conjugation acts trivially. So θ_* acts identically with the same argument showing that $\theta_* = \mathrm{id}$ on $H_*(GLA, M)$ for M any (GLA/EA)-module, and meanwhile

$$H_*(GLA, M) = H_*(BGLA, M) = H_*(BGLA^+, M).$$

It follows that θ induces an isomorphism on $BGLA^+$ for all kinds of homology, and so it is a homotopy equivalence.

Step 2 (Exercise) Any self-homomorphism of the monoid M of all embeddings $\theta : \mathbb{N} \to \mathbb{N}$ is trivial.

This proves the lemma. $\square$

Proposition $BGLA^+$ *is a homotopy commutative associative H-space.*

Proof Choose $\mathbb{N} \coprod \mathbb{N} \hookrightarrow \mathbb{N}$. This gives a homomorphism $GLA \times GLA \to GLA$ and hence a map $B(GLA \times GLA)^+ \to BGL\,A^+$. But

$$B(GLA \times GLA)^+ = (BGLA \times BGLA)^+ = BGLA^+ \times BGLA^+,$$

which gives a product. The previous lemma shows independence of the choice of $\mathbb{N} \coprod \mathbb{N} \hookrightarrow \mathbb{N}$. Also, the restriction of the product to both of $* \times BGLA^+$ and $BGLA^+ \times *$ are homotopic to the identity by the lemma again. We thus have an H-space. $\square$

Example For $F = \bar{\mathbb{F}}_p$, compute $\pi_* BGLF^+ = K_* F$.

To this end, following [R. Brauer], construct a map $BG \to BU$ using character theory starting from a representation of a finite group G over F. Namely, suppose given $G \xrightarrow{\rho} GL_n F$. Then choose an embedding

$$F^* \subset \{\text{roots of unity}\} \subset \mathbb{C},$$

where $F^* = \coprod_{\ell \neq p} \mathbb{Q}_\ell/\mathbb{Z}_\ell$. Define the *Brauer character* χ to be

$$\chi(\rho)(g) = \text{the sum of the eigenvalues of } \rho(g) \text{ in } \mathbb{C}^*.$$

Theorem (Brauer) $\chi(\rho)$ *is the character of a virtual representation of G over $\mathbb{C}$, that is*

$$\chi(\rho) = \chi V_1 - \chi V_2.$$

We get $BGLF \to BU$ hence $BGLF^+ \to BU$ since BU is simply connected, and this is almost a homology equivalence.

****

As noted in the Foreword, the class and the notes ended abruptly and embarrassedly here with this vague remark on homology equivalence owing to the end of class time, which Quillen always meticulously respected. The precise statement proved in Quillen's 1972 paper [9] is that this map lifts to the homotopy fiber of $\Psi^q - 1$, where Ψ is an Adams operation, and this lift is almost a homotopy equivalence.

Afterword by Mikhail Karpranov

This volume contains the notes of a course given by Daniel Quillen at MIT in 1979/80, almost 40 years ago. The goal of the course was to provide an introduction to Quillen's fundamental work on higher algebraic K-theory. It begins from scratch, making very minimal requirements on the mathematical background of the participants (only an elementary knowledge of groups, rings etc. is assumed), and in the space of 38 lectures brings them to the subject matter of Quillen's original work. Among the subjects covered are:

- Group cohomology.
- Homological algebra. Derived categories.
- Simplicial sets and simplicial homotopy theory.
- Homotopy theory of categories via classifying spaces.
- Higher K-theory via Quillen's Q-construction.
- Sketches of relations to other techniques:
  - Tits complexes (used in Quillen's proof of the finite generation of K-groups for algebraic integers and curves over finite fields).
  - The definition of higher K-groups via the Plus construction.

Most of this material is available in modern textbooks and in Quillen's original and excellently written papers [9–11]. For further study the reader will definitely need to master these as well. Among the standard modern references on homological algebra one can name the books of Gelfand–Manin [2] and Weibel [15] as well as the classic [1] of Cartan–Eilenberg. The basics of simplicial homotopy theory can also be found in Gelfand–Manin [2] and a more systematic treatment in the book of Goerss–Jardine [3]. The books of Srinivas [13] and Weibel [16] provide systematic treatments of various approaches to higher K-theory.

The present volume gives a shortened overview of the subject which may be useful and less intimidating for the beginner than a more fundamental course of reading which can come later.

Comparison of modern treatments shows how difficult it is to improve on Quillen. The sheer perfection of his work is almost frightening. As far as the foundations of

© Springer Nature Switzerland AG 2020

R. Penner, *Topology and K-Theory*, Lecture Notes in Mathematics 2262,
https://doi.org/10.1007/978-3-030-43996-5

K-theory go, one important later development is the S-construction of Segal and Waldhausen [14, 16] which can be viewed as a somewhat more flexible analog of the Q-construction, applicable to a wider class of categories. Quillen famously expressed hope [10] that the techniques around his theorems A and B will some day be incorporated into a general homotopy theory for toposes. One can see Lurie's theory of ∞-categories and ∞-toposes [7] as a confirmation of this hope.

These notes, with their informal style, provide a direct window into the way of operation of a great mind. Their preservation, owing to the effort of Robert Penner, will be very much appreciated by the readers.

References

1. H. Cartan, S. Eilenberg, *Homological Algebra* (Princeton University Press, Princeton, 1956)
2. S.I. Gelfand, Y.I. Manin, *Methods of Homological Algebra* (Springer, Berlin, 1997)
3. P.G. Goerss, J.F. Jardine, *Simplicial Homotopy Theory* (Birkhäuser, Basel, 1999)
4. D. Grayson, Higher algebraic K-theory II (after Daniel Quillen). *Lecture Notes in Mathematics*, vol. 551 (Springer, Berlin, 1976), pp. 217–240
5. D. Grayson, *Finite generation of K-groups of a curve over a finite field (after Daniel Quillen)*, vol. 966, Lecture Notes in Mathematics (Springer, Berlin, 1982), pp. 69–90
6. A. Grothendieck, J.-L. Verdier, *Préfaisceaux (SGA4 Éxp. 1)*, Lecture Notes in Mathematics, vol. 269 (Springer, Berlin, 1972), pp. 17–233
7. J. Lurie, *Higher Topos Theory* (Princeton University Press, Princeton, 2009)
8. J. Milnor, *Introduction to Algebraic K-Theory, Annals of Mathematics Studies*, vol. 72 (Princeton University Press, Princeton, 1971)
9. D. Quillen, On the cohomology and K-theory of the general linear groups over a finite field. Ann. Math. **96**, 552–586 (1972)
10. D. Quillen, *Algebraic K-theory I*, vol. 341, Lecture Notes in Mathematics (Springer, Berlin, 1973), pp. 85–147
11. D. Quillen, *Finite generation of the groups K_i of rings of algebraic integers*, vol. 341, Lecture Notes in Mathematics (Springer, Berlin, 1973), pp. 179–198
12. J.-P. Serre, *Local Fields* (Springer, Berlin, 1979)
13. V. Srinivas, *Algebraic K-Theory* (Birkäuser, Basel, 1996)
14. F. Waldhausen, Algebraic K-theory of generalized free products. Ann. Math. **108**, 135–204 (1978)
15. C.A. Weibel, *Introduction to Homological Algebra* (Cambridge University Press, Cambridge, 1994)
16. C.A. Weibel, *The K-book: An Introduction to Algebraic K-theory* (American Mathematical Society Publishing, Providence, 2013)

© Springer Nature Switzerland AG 2020   209
R. Penner, *Topology and K-Theory*, Lecture Notes in Mathematics 2262,
https://doi.org/10.1007/978-3-030-43996-5

Index

A
Abelian category, 51, 55
Acyclic map, 201
Additive category, 54
Additive functor, 23
Additive non-abelian category, 57
Adjoint functors, 43
Adjunction formula, 127
Admissible layer, 172
Admissible morphism, 144
Arrow, 6
Axioms for simplicial objects, 10

B
Bi-simplicial set, 179
Burnside ring, 103

C
Cartesian morphism, 63
Category, 6
Category object, 156
Classifying space, 156
1-coboundary, 6
2-cocycle, 3
Co-chain complex, 4
Cofibered category, 64
Cogroup structure, 52
Cohomology of cyclic groups, 28
Cohomology of free groups, 22
Cokernel, 54
Complex in abelian category, 72
Conical contractibility, 17
Conical contraction, 170
Connecting homomorphism, 23

Contractibility, 169
Crossed homomorphism, 5

D
δ-functor, 23
Derivation, 5, 22
Derived category, 91
Devissage Theorem, 106
Direct limit, 42
Direct sum, 42
Discrete category, 47, 69
Dold–Thom Theorem, 182

E
Effaceable functor, 21
Equalizer, 41
Exact category, 143
Exact sequence of K-groups, 191

F
Fiber, 47
Fibered category, 63
Final object, 40
First homotopy property, 95
Full subcategory, 73
Functor, 7

G
Generalized Kan formula, 134
Geometric realization, 155
Grothendieck group, 102
Group completion, 101

© Springer Nature Switzerland AG 2020
R. Penner, *Topology and K-Theory*, Lecture Notes in Mathematics 2262,
https://doi.org/10.1007/978-3-030-43996-5

Group extension, 1
Groupoid, 7
G-set, 41

H
Higher K-groups, 161
Hochschild–Serre spectral sequence, 140
Homology groups with coefficients in a module, 13
Homology of category with values in a functor, 12
Homology of cyclic groups, 27
Homology of simplicial complex, 16
Homotopy between functors, 72
Homotopy equivalence, 169
Hyper-homology spectral sequence, 126

I
Increasing filtration, 123
Initial object, 40
Injective object, 71

K
Kan formula, 47
Kernel, 55

L
Layer, 170
Left-derived functor, 91
Left-fiber, 47
Leray spectral sequence, 133
Localization, 101
Localization Theorem, 191

M
Mackey formula, 35
Mapping cone, 87
Mapping cylinder, 85
Modules, 2
Monoid, 6
Morphisms, 6

N
Nerve of a category, 7
Nerve of topological category, 156
Normalization Theorem, 15

O
Objects, 6

P
Plus construction, 201
Poset, 6
Postnikov filtration, 123
Pre-cofibered category, 64
Pre-fibered category, 63
Projective object, 71
Projective resolution, 73
Puppe sequence, 111

Q
Quasi-fibration, 182
Quasi-isomorphism, 72
Quis, 72

R
Representable functor, 40
Resolution of a complex, 73
Resolution Theorem, 107, 144, 196
Right fiber, 47

S
Schreier Lemma, 106
Schur multiplier, 203
Schur–Zassenhaus Theorem, 36
Second cohomology group, 4
Section, 3
Semi-simplicial space, 155
Serpent Lemma, 59
Simplex in a category, 7
Simplicial abelian group, 10
Simplicial object, 9
Simplicial set, 7
Small category, 6
Spectral sequence, 115
Split extension, 22
s.s., 155
Steinberg group, 203
Steinberg module, 177

T
Theorem A, 189
Theorem B, 191
Thick subcategory, 147
Tits building, 177
Tits complex, 187

Topological category, 156
Transfer map, 33

U
Universal central extension, 203

Y
Yoneda Lemma, 39

Printed in the United States
By Bookmasters